Advances in Computation and Modeling of Materials Mechanics

Advances in Computation and Modeling of Materials Mechanics

Guest Editors

Hai Huang
Abduljabar Alsayoud
Yucheng Lan

MDPI

Basel • Beijing • Wuhan • Barcelona • Belgrade • Novi Sad • Cluj • Manchester

Guest Editors
Hai Huang
Key Laboratory of Material Physics of the Ministry of Education
Zhengzhou University
Zhengzhou
China

Abduljabar Alsayoud
Mechanical Engineering Department
King Fahd University of Petroleum & Minerals
Dhahran
Saudi Arabia

Yucheng Lan
Department of Physics and Engineering Physics
Morgan State University
Baltimore
USA

Editorial Office
MDPI AG
Grosspeteranlage 5
4052 Basel, Switzerland

This is a reprint of the Special Issue, published open access by the journal *Materials* (ISSN 1996-1944), freely accessible at: https://www.mdpi.com/journal/materials/special_issues/4P3JU6W08K.

For citation purposes, cite each article independently as indicated on the article page online and as indicated below:

Lastname, A.A.; Lastname, B.B. Article Title. *Journal Name* **Year**, *Volume Number*, Page Range.

ISBN 978-3-7258-6061-6 (Hbk)
ISBN 978-3-7258-6062-3 (PDF)
https://doi.org/10.3390/books978-3-7258-6062-3

Contents

About the Editors vii

Preface ix

Xu Yu, Hanlong Wang and Hai Huang
Investigating Helium-Induced Thermal Conductivity Degradation in Fusion-Relevant Copper: A Molecular Dynamics Approach
Reprinted from: *Materials* **2025**, *18*, 3702, https://doi.org/10.3390/ma18153702 1

Yu Gu, Jiayue Wang, Jun Zhang, Yu Zhang, Bushi Dai, Yu Li, Guangchao Liu, Li Bao and Rihuan Lu
Machine Learning-Driven Design and Optimization of Multi-Metal Nitride Hard Coatings via Multi-Arc Ion Plating Using Genetic Algorithm and Support Vector Regression
Reprinted from: *Materials* **2025**, *18*, 3478, https://doi.org/10.3390/ma18153478 13

Yanan Cui, Jingjing Sun, Meng Li and Bingsheng Li
First-Principles Calculations of Hydrogen Solution and Diffusion in 3C-SiC Grain Boundaries
Reprinted from: *Materials* **2025**, *18*, 2118, https://doi.org/10.3390/ma18092118 42

Xiaoxin Ge, Yanxin Jiang, Xu Yu, Guopeng Zhang, Yunjia Shi, Bin Cai, et al.
Preparation and Characterization of Graphene-Nanosheet-Reinforced Ni-17Mo Alloy Composites for Advanced Nuclear Reactor Applications
Reprinted from: *Materials* **2025**, *18*, 1061, https://doi.org/10.3390/ma18051061 55

Ling Xiao, Yaxin Yao, Shuai Chen, Mengting Lai and Guanghong Zhu
Analytical Model of Temperature-Induced Deformation for Tunable Thermal Expansion Metamaterial
Reprinted from: *Materials* **2025**, *18*, 532, https://doi.org/10.3390/ma18030532 68

Qing Peng, Gen Chen, Zeyu Huang, Xue Chen, Ao Li, Xintian Cai, et al.
Molecular Dynamics Insights into Mechanical Stability, Elastic Properties, and Fracture Behavior of PHOTH-Graphene
Reprinted from: *Materials* **2024**, *17*, 4740, https://doi.org/10.3390/ma17194740 82

Yanqing Li, Ziming Zhang, Junyi Cheng, Zhaofeng Liu, Chao Yin, Chao Wang and Jianzheng Guo
The Integration of ANN and FEA and Its Application to Property Prediction of Dual-Performance Turbine Disks
Reprinted from: *Materials* **2024**, *17*, 3045, https://doi.org/10.3390/ma17133045 102

Changyin Wang, Jianyao Yao, Xu Zhang, Yulin Wu, Xuyang Liu, Hao Liu, et al.
Fatigue Life Data Fusion Method of Different Stress Ratios Based on Strain Energy Density
Reprinted from: *Materials* **2024**, *17*, 2982, https://doi.org/10.3390/ma17122982 122

Kehao Qiang, Shisong Wang, Haowen Wang, Zhulin Zeng and Liangzhao Qi
Study on the Constitutive Modeling of (2.5 vol%TiB + 2.5 vol%TiC)/TC4 Composites under Hot Compression Conditions
Reprinted from: *Materials* **2024**, *17*, 619, https://doi.org/10.3390/ma17030619 148

About the Editors

Hai Huang

Hai Huang is an Associate Professor and Master's Supervisor at the School of Physics, Zhengzhou University, China. His interdisciplinary research focuses on nuclear technology and advanced materials, with expertise in radiation-resistant materials for nuclear reactors, structural materials for renewable energy systems, and multi-scale computational modeling of materials. He earned his Ph.D. in Nuclear Technology and Materials Engineering from Nanjing University of Aeronautics and Astronautics in 2019 and was a Visiting Scholar at the University of Michigan, Ann Arbor, in 2018. He has been recognized for his academic contributions through appointments such as the Henan Provincial Young Talent Support Program and Zhengzhou University Elite Doctoral Talent Program. He has led multiple research projects, including grants from the National Natural Science Foundation of China and Key Projects for Science and Technology Development of Henan Province. His work has resulted in 50+ publications, 8 patents, and 8 software copyrights. He also actively contributes to academia as a Guest Editor for *Nanomaterials* and *Materials*, an Editorial Board Member of TUNGSTEN, and a reviewer for leading journals such as the *Journal of Nuclear Materials* and *Computational Materials Science*. His scientific pursuits continue to advance materials science for nuclear and renewable energy applications.

Abduljabar Alsayoud

Abduljabar Alsayoud is an Assistant Professor in the Department of Materials Science and Engineering, where he specializes in computational materials design using density functional theory (DFT) and molecular dynamics (MD). His educational background includes a Ph.D. in Materials Science and Engineering from the University of Arizona (2017), an M.Sc. from King Abdullah University of Science and Technology (2010), and a B.Sc. in Mechanical Engineering from KFUPM (2008). His research focuses on addressing challenges in hydrogen and energy storage, particularly through the development of advanced materials for hydrogen separation membranes and nano-engineered carbon-based materials. Currently, he is affiliated with the Interdisciplinary Research Center for Hydrogen and Energy Storage at King Fahd University of Petroleum and Minerals (KFUPM), where he leads projects on metallic membrane design for hydrogen purification. His scientific contributions include pioneering work on molten salts for solar thermal storage, combining Raman spectroscopy with DFT modeling, and developing charge transfer potentials for MD simulations. He has also investigated the thermal conductivity of 2D fullerene materials for metamaterials and the thermoelectric performance of doped half-Heusler alloys. In addition to research, he teaches courses such as Atomistic Simulation of Materials, Thermodynamics of Materials, and Integrated Computational Materials Science, fostering the next generation of materials scientists. His work bridges theoretical simulations with practical applications, aiming to advance sustainable energy technologies.

Yucheng Lan

Yucheng Lan is a Professor of Physics at Morgan State University, specializing in experimental condensed matter physics and materials science. He holds a distinguished academic record, having earned his Ph.D. in experimental condensed matter physics from the Institute of Physics at the Chinese Academy of Sciences, following bachelor's and master's degrees in solid-state physics from Jilin University. With extensive research experience spanning multiple prestigious institutions, including the University of Houston, Boston College, and the Chinese Academy of Sciences, he has established himself as a leading expert in renewable energy materials and structural characterization techniques. His current research focuses on three key areas: 1) renewable clean energy materials (thermoelectrics, photovoltaics, and photo-/electro-catalytics), 2) sensing technologies, and 3) advanced structural characterization methods using X-ray diffraction and electron microscopy. As Principal Investigator or co-PI, he has led numerous projects funded by the National Science Foundation, Department of Energy, Department of Defense, and various national laboratories and industrial partners. His professional journey includes appointments as a visiting professor at Delaware State University and an ORAU Fellow at the US Army Research Laboratory. An exceptionally productive scholar, he has authored over 150 peer-reviewed publications (h-index 46, citations exceeding 17,000), one monograph, 12 book chapters, and holds more than 10 patents. He serves as Task Chair for the International Centre for Diffraction Data (ICDD) and has guest-edited two special journal issues. His work bridges fundamental materials research with practical applications in energy conversion and sensing technologies, making significant contributions to the development of sustainable energy solutions through advanced materials design and characterization.

Preface

It is our privilege to present this Reprint compiling the Special Issue "Advances in Computation and Modeling of Materials Mechanics" published in *Materials*. This collection showcases cutting-edge research where computational methodologies intersect with materials mechanics, addressing critical challenges in composite characterization, irradiation effects in nuclear materials, multi-scale mechanical simulations, and machine learning-driven material design.

The motivation for this Special Issue emerged from the rapidly evolving role of computational techniques in decoding complex material behaviors under extreme conditions—knowledge essential for next-generation material development. By bridging simulation frameworks with experimental validation, these studies collectively advance predictive capabilities for material performance.

This Reprint serves researchers and engineers in nuclear engineering, computational materials science, mechanics, and related interdisciplinary fields. We extend profound gratitude to all contributing authors for their scientific excellence and to the peer reviewers whose rigorous evaluations strengthened this work. We express special thanks to the *Materials* editorial team for their unwavering support throughout this scholarly endeavor.

Hai Huang, Abduljabar Alsayoud, and Yucheng Lan
Guest Editors

materials

MDPI

Article

Investigating Helium-Induced Thermal Conductivity Degradation in Fusion-Relevant Copper: A Molecular Dynamics Approach

Xu Yu [1], Hanlong Wang [2] and Hai Huang [1,*]

1 Key Laboratory of Material Physics, Ministry of Education, School of Physics, Zhengzhou University, Zhengzhou 450001, China
2 International Joint Laboratory for Integrated Circuits Design and Application, Ministry of Education, School of Physics, Zhengzhou University, Zhengzhou 450001, China
* Correspondence: huanghai@zzu.edu.cn

Abstract: Copper alloys are critical heat sink materials for fusion reactor divertors due to their high thermal conductivity (TC) and strength, yet their performance under extreme particle bombardment and heat fluxes in future tokamaks requires enhancement. While neutron-induced transmutation helium affects the properties of copper, the atomistic mechanisms linking helium bubble size to thermal transport remain unclear. This study employs non-equilibrium molecular dynamics (NEMD) simulations to isolate the effect of bubble diameter (10, 20, 30, 40 Å) on TC in copper, maintaining a constant He-to-vacancy ratio of 2.5. Results demonstrate that larger bubbles significantly impair TC. This reduction correlates with increased Kapitza thermal resistance and pronounced lattice distortion from outward helium diffusion, intensifying phonon scattering. Phonon density of states (PDOS) analysis reveals diminished low-frequency peaks and an elevated high-frequency peak for bubbles >30 Å, confirming phonon confinement and localized vibrational modes. The PDOS overlap factor decreases with bubble size, directly linking microstructural evolution to thermal resistance. These findings elucidate the size-dependent mechanisms of helium bubble impacts on thermal transport in copper divertor materials.

Keywords: copper alloys; thermal transport; helium bubble; fusion reactor divertor; non-equilibrium molecular dynamics

1. Introduction

Copper (Cu) and its alloys are prominent candidates for heat sink applications in fusion reactor divertors owing to their exceptional strength, superior thermal conductivity (TC), and high thermal stability [1–3]. However, these plasma-facing components within advanced magnetic confinement fusion systems must withstand extreme operational environments characterized by intense particle bombardment and substantial thermal fluxes [4–6]. In particular, future Tokamak designs, such as CFETR and DEMO, will impose even greater demands on divertor materials, requiring resilience against heat loads potentially reaching 20 $MW{\cdot}m^{-2}$ during extended steady-state operation [7,8]. Consequently, the thermophysical performance of Cu-based alloys under irradiation necessitates significant enhancement to meet these challenges.

High-dose neutron irradiation generates point defects, dislocation loops, and stacking-fault tetrahedra (SFTs) within materials, while simultaneously producing gaseous transmutation products (e.g., He atoms) through nuclear reaction [9–11]. These irradiation-induced alterations significantly degrade the thermal properties of copper and its alloys, ultimately shortening divertor service life [12,13]. For example, research into irradiation's impact on thermal transport includes Zhang et al.'s investigation, which revealed that vacancies diminish the phonon TC of single-crystal copper via phonon scattering, with TC reduction proportional to vacancy concentration [14]. Ye et al. [15] demonstrated that tungsten doping (4 at.%) drastically reduces the TC of copper by approximately 80%, attributing this effect to significant differences in the electronic states of W and Cu atoms compared to vacancies or self-interstitial atoms, which impede electron transport. Furthermore, Fabritsiev et al. [16,17] observed a 25% decrease in TC for both CuCrZr and GlidCop Al-25 alloys at 573 K under neutron irradiation (2 dpa), identifying the nuclear transmutation effect as the primary contributor—responsible for 80% of this reduction. Nevertheless, the fundamental atomistic mechanisms through which irradiation damage, particularly neutron-induced transmutation helium, governs thermal transport degradation in Cu-based materials remain inadequately characterized.

Molecular dynamics (MD) simulations compute macroscopic system properties by statistically analyzing atomic positions and momenta within frameworks derived from statistical mechanics [18,19]. This approach can provide atomic-scale insights into the fundamental mechanisms by which irradiation damage degrades materials' thermal transport characteristics [20,21]. Among the available computational techniques, the non-equilibrium MD (NEMD) method has become a principal tool for simulating thermal transport in nanomaterials and has yielded substantial advances in understanding thermal transport within material systems containing transmutation-produced He atoms [22,23]. For example, Petersson et al. [24] demonstrated that even low concentrations of interstitial helium in tungsten markedly reduce TC; however, vacancy–helium interactions that trap helium atoms can mitigate this degradation by reducing lattice distortion. Conversely, Sharma et al. [25] indicated that within nickel, increasing helium bubble diameter and internal helium concentration elevate both the temperature gradient and Kapitza resistance across the bubble region. Helium demonstrates negligible solubility in metallic systems, driving its precipitation into nanoscale bubbles within nuclear structural materials [26,27]. Although prior research has explored the influence of transmutation-produced helium on the structural, mechanical, and thermal properties of copper and its alloys [28,29], the atomistic mechanisms through which helium bubble dimensions govern thermal transport remain unresolved.

To address this knowledge gap, in this work, the NEMD method is utilized to examine the thermal transport behavior in copper materials from an atomistic perspective, with an emphasis on elucidating how helium bubbles impact thermal conductivity. We thoroughly investigate how helium bubble size affects the TC of copper materials. It is noteworthy that the He-to-vacancy ratio within bubbles of varying sizes remains constant, which allows us to isolate the effect of bubble size on thermal transport. This methodological approach aligns with established practices in the field, as evidenced by numerous prior studies [30,31]. The results highlight that larger helium bubbles more significantly diminish the TC of copper materials.

2. Simulation Methodology

All the MD simulations were conducted using the Large-scale Atomic/Molecular Massively Parallel Simulator code (LAMMPS, version 29 Aug 2024) [32], with atomic configurations visualized via the Open Visualization Tool software (OVITO, version 3.11.3) [33]. For modeling the Cu–He system, we adopted the well-established embedded atom method (EAM) potential developed by Kashinath et al. [34], which accounts for Cu–Cu, He–He, and Cu–He interatomic interactions. This potential reliably captures helium diffusion behavior and provides accurate predictions of both migration and formation energies for helium in face-centered cubic (fcc) copper [34].

The simulation setup began with the creation of a single-crystal copper structure with dimensions of $150a_0 \times 15a_0 \times 15a_0$ (where $a_0 = 3.615$ Å represents the Cu lattice constant [35,36]), containing approximately 135,000 atoms (see Figure 1a). The crystallographic orientations were aligned with the simulation axes, assigning [100], [010], and [001] directions to x, y, and z coordinates, respectively. A spherical void was then introduced at the system's center, into which He atoms were randomly distributed to create an artificial bubble structure (see Figure 1b). Stress relaxation was performed via conjugate gradient energy minimization under zero external pressure. Periodic boundaries were maintained in all directions throughout simulations. Before thermal transport analysis, the system underwent 0.1 ns of NPT ensemble equilibration at 300 K and zero pressure using a Nosé–Hoover thermostat–barostat to achieve stability. Four distinct bubble diameters (10, 20, 30, and 40 Å) were investigated to assess size effects on thermal properties, maintaining a constant He-to-vacancy ratio of 2.5 [30]. For baseline comparison, a defect-free copper crystal was simulated under identical conditions.

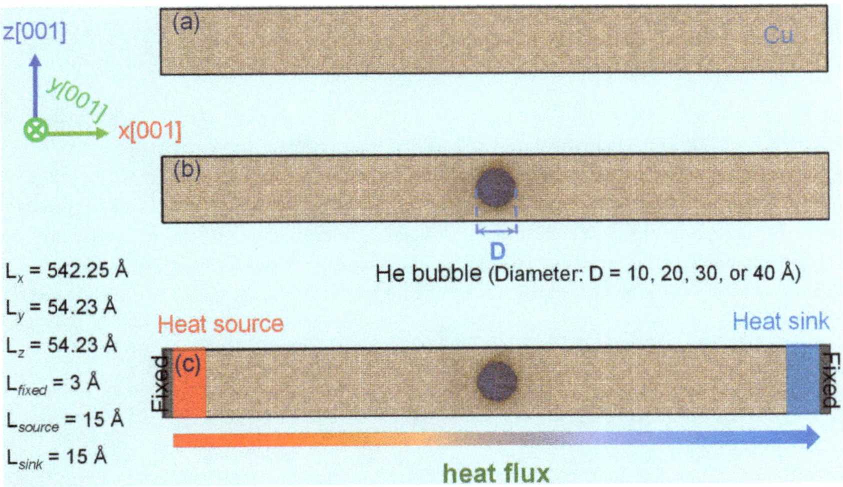

Figure 1. NEMD computational framework employed to analyze thermal transport in a copper system embedded with helium bubbles. (**a**) Atomic configuration of a pristine single-crystal copper structure. (**b**) Corresponding atomic model incorporating a helium bubble. (**c**) Schematic representation of the simulation domain, explicitly identifying the fixed boundary layers, heat source region, and heat sink region.

To establish a thermal gradient, periodicity along the *x*-axis was disrupted by fixing outer layers (3.0 Å thickness) at both ends of the simulation cell. Adjacent to these fixed regions, paired 15.0 Å thick reservoirs generated a constant heat flux density of 5.5×10^9 W·m^{-2} (see Figure 1c). Heat injection occurred at the left reservoir (source), propagated unidirectionally through the system, and dissipated at the right reservoir (sink). Atoms within the central adiabatic region evolved under NVE ensemble dynamics, while Langevin thermostats regulated reservoir temperatures. Following 0.2 ns of equilibration for system stabilization, temperature data were collected over 4 ns for time-averaging. Local temperatures were recorded within 12.7 Å thick bins along the *x*-axis to construct temperature profiles, enabling thermal transport efficiency evaluation in helium-bubble-containing copper. All simulations employed this consistent heat flux density methodology. The chosen flux magnitude (10^9–10^{10} W·m^{-2}) aligns with established NEMD methodologies for simulating extreme divertor thermal loads in fusion reactors [37,38]. Note that the elevated flux magnitude—exceeding physical conditions—intentionally enhances the temperature gradient resolution for analysis, consistent with comparable computational studies [37–39]. To mitigate stochastic uncertainties in defect configurations, each case was replicated three times, and the averaged results were reported.

3. Results and Discussion

To investigate the influence of helium bubbles, we first examined the thermal transport behavior of defect-free copper as a reference baseline. In the course of NEMD simulations, temperature field data were captured at consistent intervals of 0.5 ns to monitor the progression of thermal transport within a pristine copper crystal. Figure 2a illustrates that, at the 0.5 ns mark, there was a pronounced rise in temperature adjacent to the heat source, accompanied by a decline near the heat sink. Nonetheless, the temperature profile revealed substantial nonlinearity before 0.5 ns, suggesting that steady-state thermal transport had not been attained. After 1 ns, the temperature distribution achieved stability, manifesting an almost linear gradient along the x-axis. The temperature gradient was determined by fitting the temperature data at the 4 ns timepoint and was subsequently integrated with Fourier's law [40]

$$J = \kappa \frac{dT}{dx} \tag{1}$$

where J denotes the heat flux density, κ represents the TC, and dT/dx is the temperature gradient. Calculations yielded a TC of 7.98 $W \cdot m^{-1} \cdot K^{-1}$ for the defect-free copper crystal at 300 K, closely aligning with the 7.86 $W \cdot m^{-1} \cdot K^{-1}$ documented in a previous study [35] and thereby confirming the accuracy of the computational approach employed in this research. Typically, insights into the mechanisms of thermal transport can be gained indirectly by examining the phonon density of states (PDOS), as this provides a quantitative measure of variations in the phonon spectrum [41–43]. The PDOS is determined by applying the fast Fourier transform to the atomic velocity autocorrelation function (VACF) obtained from MD trajectories, as expressed in the equation [44,45]

$$PDOS(\omega) = \int_{-\infty}^{\infty} e^{-i\omega t} VACF(t) dt \tag{2}$$

where

$$VACF(t) = \frac{1}{N} \sum_{j=1}^{N} \langle v_j(0) v_j(t) \rangle \tag{3}$$

In this formulation, ω represents the phonon vibration frequency, N is the number of atoms within the analysis domain, $v_j(t)$ denotes the velocity vector of the jth atom at time t, and <···> signifies the ensemble average. Consequently, the PDOS characteristics of the defect-free copper at 300 K were computed and are presented in Figure 2b. The analysis reveals that in defect-free coppers at 300 K, low-frequency phonon modes, predominantly distributed between 0 and 10 THz, dominate the PDOS spectrum. Spectral decomposition reveals a bimodal distribution featuring a more intense primary peak at 4.6 THz compared to the secondary peak at 7.0 THz. These findings are consistent with those reported in prior studies [39,46], thereby further validating the reliability of the simulation methodology employed in this work.

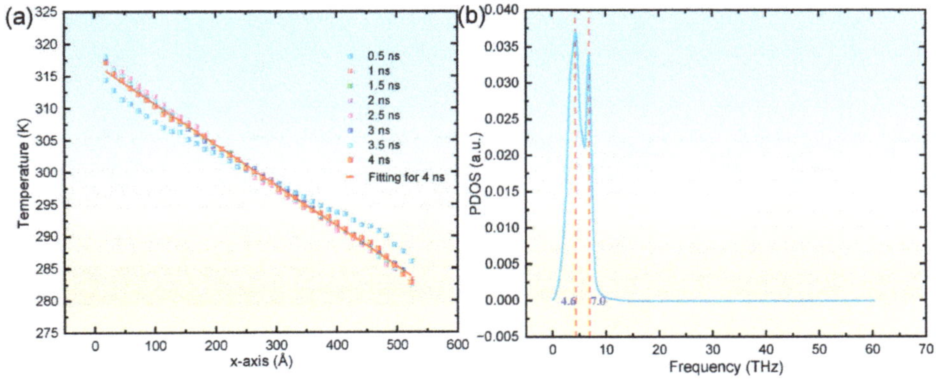

Figure 2. Temporal evolution of the spatial temperature distribution along the x-axis for a pristine single-crystal copper system at 300 K (**a**), alongside its corresponding PDOS spectrum at 4.0 ns (**b**).

Numerous investigations have established that examining alterations in the PDOS serves as an effective method to elucidate the mechanisms by which microstructural characteristics, such as defects and interfaces, impact thermal transport properties [41]. This area of research has witnessed considerable advancements, as evidenced by various studies [47]. Leveraging this established framework, the current study undertakes a systematic examination of how the size of helium bubbles influences the TC within copper. Subsequently, we performed a detailed analysis to assess the impact of helium bubbles of different sizes on the thermal transport characteristics of copper. Figure 3 depicts the time-dependent changes in temperature distribution throughout the heat transport process in four distinct systems that incorporate helium bubbles. During the early phase of the simulation (at t = 0.5 ns) for each system containing helium bubbles, an elevation in temperature was noted in the vicinity of the heat source, concurrent with a reduction in temperature near the heat sink. This behavior mirrors that observed in defect-free copper. It is noteworthy that, at this juncture, the temperature profiles within the regions devoid of bubbles continued to display nonlinear traits, signifying that the system had not attained a state of steady heat conduction. As the simulation advanced beyond 1 ns, the temperature distribution achieved a stable configuration. Generally, an increase in the size of helium bubbles correlates with heightened fluctuations in the temperature distribution within the respective system. Within areas lacking bubbles, the temperature exhibited a distinct linear gradient along the x-axis. Conversely, regions housing helium bubbles experienced notable decreases in temperature. A quantitative assessment indicated temperature declines of 2.943 K, 3.875 K, 9.132 K, and 14.581 K for bubble diameters measuring 10 Å, 20 Å, 30 Å, and 40 Å, respectively. These findings unequivocally illustrate that the temperature gradient intensifies significantly as the size of the helium bubbles increases.

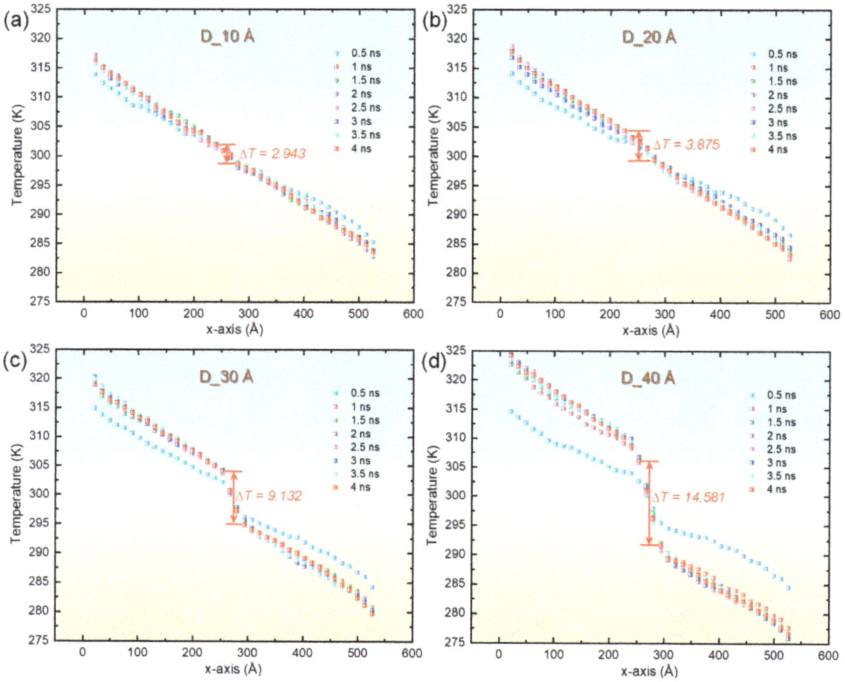

Figure 3. Temporal evolution of the spatial temperature distribution along the *x*-axis for single-crystal copper systems embedded with helium bubbles of varying diameters. (**a**) 10 Å. (**b**) 20 Å. (**c**) 30 Å. (**d**) 40 Å.

For comparative analysis, Figure 4a depicts the steady-state temperature distributions within copper systems that incorporate helium bubbles of diverse sizes. The examination reveals marked differences in the temperature gradient across the bubble regions, with the magnitude of this gradient escalating significantly as the bubble diameter increases. This effect may stem from the substantially lower thermal conductivity of the helium bubble regions (approximately 0.3 $W \cdot m^{-1} \cdot K^{-1}$ [48]) relative to the copper matrix, which induces pronounced thermal resistance effects. Additionally, as the helium bubble size expands, the temperatures adjacent to the system's heat source and heat sink exhibit increasing and decreasing trends, respectively. Furthermore, this trend becomes more accentuated with larger helium bubble sizes. This behavior may be attributed to the heightened thermal resistance associated with larger helium bubbles, which impedes heat transfer along the *x*-axis, thereby complicating the dissipation of heat near the heat source and its absorption near the heat sink. When conceptualizing the helium-bubble-containing copper as a composite material with copper as the matrix and helium as the filler particles, Maxwell's effective medium theory for spherical inclusions can be applied to determine the effective thermal conductivity [31]

$$\frac{\kappa_{eff}}{\kappa_m} = 1 + \frac{3\varphi}{\left(\frac{\frac{\kappa_1}{\kappa_m}+2}{\frac{\kappa_1}{\kappa_m}-1}\right) - \varphi} \tag{4}$$

where φ denotes the volume fraction of the void occupied by the helium bubble, and κ_m and κ_1 represent the thermal conductivities of the perfect copper crystal and the helium bubble, respectively. For a vacuum void, $\kappa_1 = 0$, whereas for a helium bubble, $\kappa_1 \approx 0.3\ W \cdot m^{-1} \cdot K^{-1}$ [48]. Given that all simulations in this study maintained a constant He-to-vacancy ratio of 2.5, the computed effective thermal conductivities for copper systems with varying bubble diameters were found to be comparable. However, as illustrated in Figure 4a, significant temperature gradients are observed between systems with bubble diameters ranging from 10 Å to 40 Å, suggesting discrepancies in effective thermal conductivity. This inconsistency arises because the Maxwell model does not account for the phonon scattering effects induced by helium atoms in the proximity of the bubbles. To more accurately characterize the influence of helium bubbles on thermal transport properties, researchers have introduced the concept of Kapitza thermal resistance (viz., the reciprocal of interfacial thermal conductance) to quantify the abrupt changes in temperature gradient near the bubble regions [25]. The Kapitza resistance is defined by

$$R = \frac{\Delta T}{J} \tag{5}$$

where ΔT signifies the temperature difference across the helium bubble region (as depicted in Figure 4a), and J represents the heat flux density along the direction of thermal transport. Figure 4b displays the Kapitza thermal resistance for copper systems containing helium bubbles of different diameters. The findings indicate a progressive increase in Kapitza resistance as the bubble diameter enlarges. Notably, the increment is relatively modest for bubbles with diameters between 10 Å and 20 Å but becomes more substantial for diameters exceeding 20 Å.

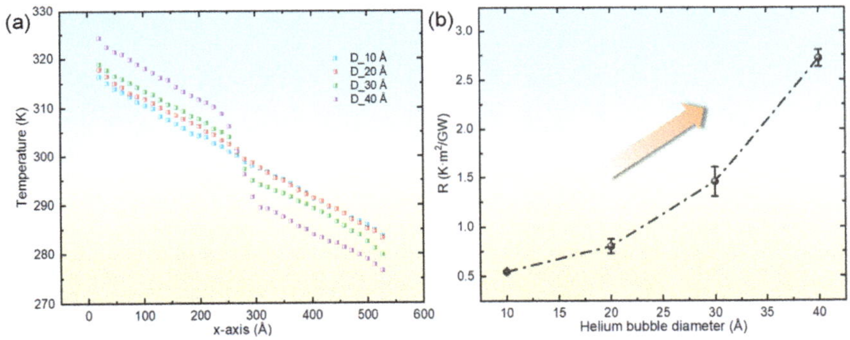

Figure 4. Instantaneous temperature distribution along the x-axis for single-crystal copper systems embedded with helium bubbles (diameters: 10–40 Å) at 4 ns (**a**), paired with the corresponding diameter-dependent Kapitza thermal resistance (**b**).

Fluctuations in Kapitza resistance are likely linked to the evolution of atomic configurations at the microscopic scale within the system. To investigate the underlying causes, we meticulously analyzed the spatial arrangement of He atoms along the x-axis in systems featuring helium bubbles with diameters ranging from 10 to 40 Å. By designating the x-coordinate of each bubble's center as the reference point, we assessed the distribution of He atoms by determining their relative abundance at various distances. Figure 5 illustrates the concentration of helium along the x-axis at both the initial (0 ns) and final (4 ns) stages of the thermal transport simulations in copper, incorporating bubbles of different sizes. A comparison of the helium distribution between the initial and final stages of the simulations reveals that larger bubbles facilitate greater outward diffusion of helium atoms, signifying their infiltration into the copper lattice [31,49]. In systems with smaller bubbles (e.g., 10 Å and 20 Å), there is negligible outward diffusion, indicating minimal distortion of the surrounding copper lattice. This observation aligns with the modest increase in Kapitza resistance depicted in Figure 4b. Conversely, for bubbles that are 30 Å in diameter, helium atoms begin to extend beyond the initial bubble boundaries, suggesting increased local lattice distortion. This phenomenon is more pronounced with a 40 Å bubble, where helium diffusion reaches approximately 23.7 Å from the center, displacing adjacent Cu atoms and inducing significant lattice deformation. Such atomic-level disruptions markedly intensify phonon scattering, thereby hindering phonon transmission and leading to a rapid escalation in Kapitza resistance. These results demonstrate a direct relationship between the size of helium bubbles and the Kapitza resistance, offering essential insights into the mechanisms governing thermal transport in copper systems containing helium bubbles.

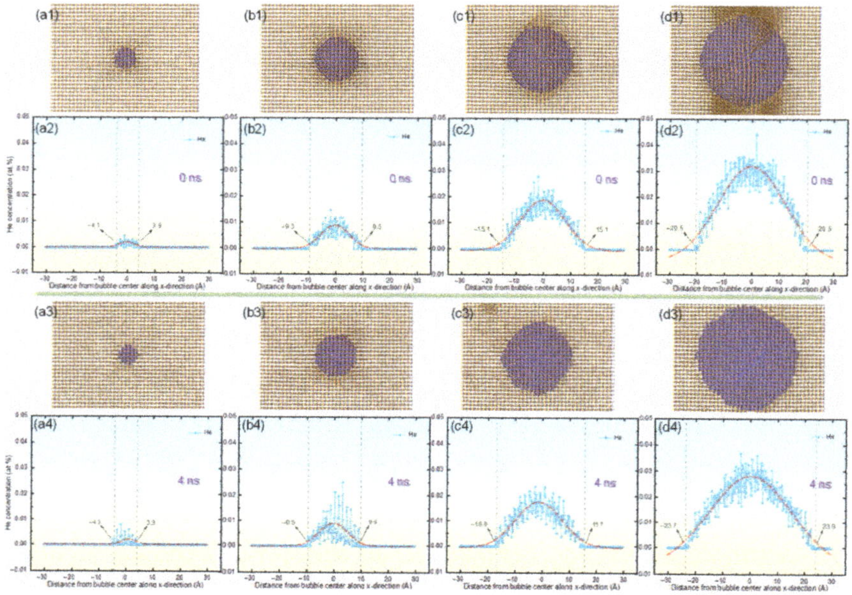

Figure 5. Evolution of helium bubble configurations and their associated concentration profiles along the x-axis during thermal transport simulations in copper, comparing initial (0 ns) and final (4 ns) states for varying bubble diameters: (**a**) 10 Å, (**b**) 20 Å, (**c**) 30 Å, and (**d**) 40 Å. Panels (**a1–d1**) and (**a3–d3**) depict the spatial distribution of helium bubbles at both timepoints, while (**a2–d2**) and (**a4–d4**) quantify the corresponding helium concentration gradients across the simulated copper matrices.

To further substantiate our analysis of helium's effect on phonon thermal transport within the crystal lattice, we computed the PDOS for each system, as illustrated in Figure 6a–d. Additionally, to quantitatively evaluate the similarity between phonon modes in systems with helium bubbles and those in pristine copper, we introduced a PDOS overlap factor [39], defined as

$$S = \int_0^{\infty} \min\{P_0(\omega), P_\mu(\omega)\} d\omega \tag{6}$$

where $P_0(\omega)$ denotes the PDOS of defect-free copper and $P_\mu(\omega)$ represents that of helium-bubble-containing systems. According to Figure 6a–d, helium bubbles with diameters of 10 Å and 20 Å cause only slight modifications to the PDOS of copper. This finding clarifies why these systems exhibit thermal conductivity akin to that of defect-free copper and maintain low Kapitza resistance. In contrast, bubbles larger than 30 Å lead to a significant drop in the low-frequency PDOS peak, coupled with a rise in the high-frequency peak. For instance, in the system with a 40 Å bubble, the low-frequency peak intensity decreases from 0.03697 to 0.03300, while the high-frequency peak intensity increases from 0.03389 to 0.03543. This shift occurs because low-frequency phonons, which are tied to long-wavelength acoustic modes, dominate thermal transport [50]. The reduction suggests that helium diffusing outward from the bubbles disturbs the copper lattice's periodicity, confining these phonons near the bubble regions and thus hindering heat conduction, which explains the elevated Kapitza resistance. Conversely, the increased high-frequency peak arises from the mass contrast between He and Cu atoms, as lighter helium impurities introduce localized high-frequency vibrational modes, boosting their density [51]. Figure 6e shows the PDOS overlap factor's variation between helium-bubble-containing systems and defect-free copper, with a declining overlap factor as bubble size grows, mirroring the trend in Kapitza resistance and affirming that phonon-based analysis effectively reveals the influence of helium bubbles on the thermal transport properties of copper.

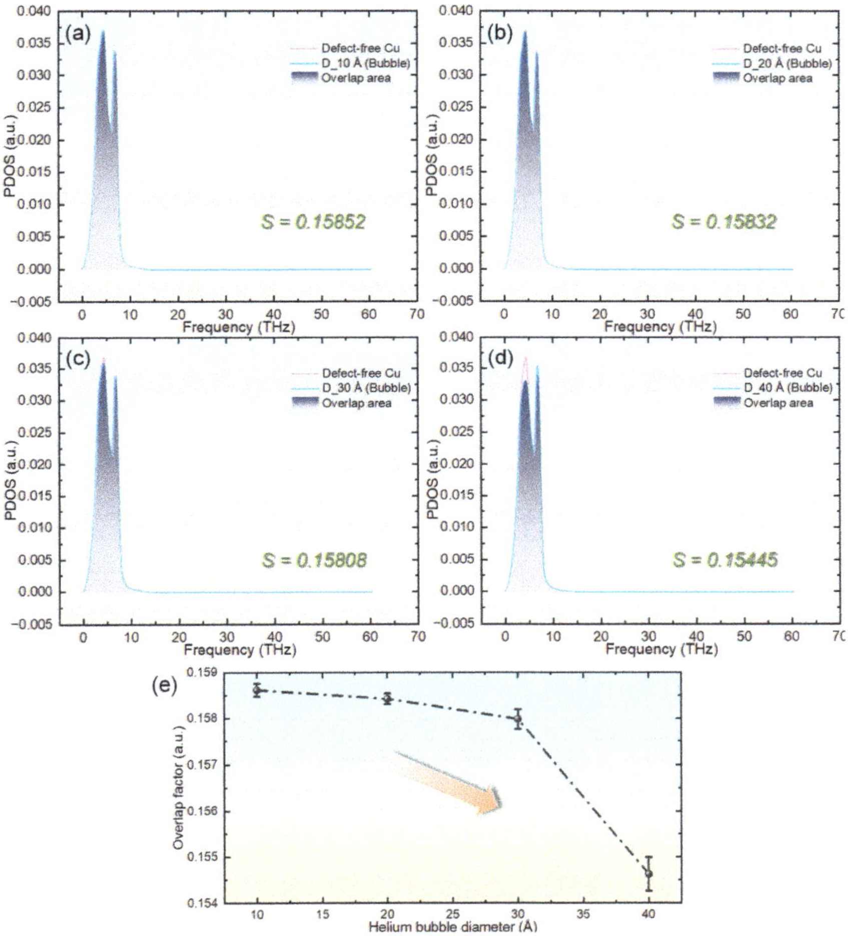

Figure 6. Comparison of PDOS spectra between pristine copper and helium-embedded copper systems with bubble diameters of (**a**) 10 Å, (**b**) 20 Å, (**c**) 30 Å, and (**d**) 40 Å. Panel (**e**) quantifies the diameter-dependent PDOS overlap factor between the two systems.

4. Conclusions

In summary, this study investigated how helium bubble size (10–40 Å diameter) influences the thermal conductivity of copper using NEMD simulations. Simulations first established a defect-free copper baseline (TC = 7.98 $W \cdot m^{-1} \cdot K^{-1}$ at 300 K), with PDOS analysis showing dominant low-frequency modes (0–10 THz), validating the methodology. The results demonstrate that larger bubbles progressively reduce TC, with temperature drops near bubbles intensifying from 2.943 K (10 Å) to 14.581 K (40 Å). Steady-state analyses reveal significantly amplified temperature gradients across bubble regions as diameters increase, attributed to helium's low intrinsic conductivity (~0.3 $W \cdot m^{-1} \cdot K^{-1}$), which induces interfacial thermal resistance. Kapitza resistance exhibits nonlinear growth, rising modestly below 20 Å but substantially above 30 Å. This trend correlates with helium diffusion patterns: bubbles >30 Å show pronounced outward helium migration (reaching 23.7 Å from the center for 40 Å bubbles), displacing Cu atoms and distorting the lattice, thereby intensifying phonon scattering. PDOS analysis confirms bubbles >30 Å suppress low-frequency phonons (critical for heat conduction) by up to approximately 11%, while enhancing high-frequency modes due to mass contrast effects. Correspondingly, the PDOS overlap factor with pristine copper decreases by 0.13–2.57% for larger bubbles, directly mirroring the trends for Kapitza resistance. These findings establish that size-dependent helium diffusion drives lattice disruption, phonon confinement, and interfacial resistance, providing atomistic insights into thermal degradation mechanisms in copper alloys for fusion applications.

Author Contributions: X.Y. and H.W. performed the atomistic simulations, helped in data analyses, and drafted the paper. H.H. improved the paper and designed and supervised the project. All authors have read and agreed to the published version of the manuscript.

Funding: This work was supported by the Young Talent Support Program of Henan Association for Science and Technology (Grant No. 2025HYTP047), the Key Project for Science and Technology Development of Henan Province (Grant No. 242102230052), and the Henan Province Postdoctoral Science Foundation (Grant No. 202102012).

Institutional Review Board Statement: Not applicable.

Informed Consent Statement: Not applicable.

Data Availability Statement: The data that support the findings of this study are available from the corresponding author upon reasonable request.

Acknowledgments: We thank the computing resources provided by National Supercomputing Center in Zhengzhou.

Conflicts of Interest: The authors declare no conflict of interest.

References

1. Kalinin, G.; Barabash, V.; Cardella, A.; Dietz, J.; Ioki, K.; Matera, R.; Santoro, R.T.; Tivey, R. Assessment and Selection of Materials for ITER In-Vessel Components. *J. Nucl. Mater.* **2000**, *283*, 10–19. [CrossRef]
2. Federici, G.; Biel, W.; Gilbert, M.R.; Kemp, R.; Taylor, N.; Wenninger, R. European DEMO Design Strategy and Consequences for Materials. *Nucl. Fusion* **2017**, *57*, 092002. [CrossRef]
3. Mota, F.; Palermo, I.; Laces, S.; Molla, J.; Ibarra, A. Potential Irradiation of Cu Alloys and Tungsten Samples in DONES. *Nucl. Fusion* **2017**, *57*, 126056. [CrossRef]
4. Bolt, H.; Barabash, V.; Krauss, W.; Linke, J.; Neu, R.; Suzuki, S.; Yoshida, N.; Asdex, U. Materials for the Plasma-Facing Components of Fusion Reactors. *J. Nucl. Mater.* **2004**, *329*, 66–73. [CrossRef]
5. Raffray, A.R.; Nygren, R.; Whyte, D.G.; Abdel-Khalik, S.; Doerner, R.; Escourbiac, F.; Evans, T.; Goldston, R.J.; Hoelzer, D.T.; Konishi, S.; et al. High Heat Flux Components—Readiness to Proceed from near Term Fusion Systems to Power Plants. *Fusion Eng. Des.* **2010**, *85*, 93–108. [CrossRef]
6. Liu, T.; Ge, X.; Chen, F.; Jiang, Y.; Huang, H. An Experimental Study on Corrosion Resistance of Ti35 Alloy and Its High-Fluence Hydrogen Bombardment Behavior. *J. Phys. Chem. Solids* **2025**, *201*, 112646. [CrossRef]
7. Li, J.; Wan, Y. Present State of Chinese Magnetic Fusion Development and Future Plans. *J. Fusion Energ.* **2019**, *38*, 113–124. [CrossRef]
8. Wan, Y.; Li, J.; Liu, Y.; Wang, X.; Chan, V.; Chen, C.; Duan, X.; Fu, P.; Gao, X.; Feng, K.; et al. Overview of the Present Progress and Activities on the CFETR. *Nucl. Fusion* **2017**, *57*, 102009. [CrossRef]
9. Zinkle, S.J.; Snead, L.L. Designing Radiation Resistance in Materials for Fusion Energy. *Annu. Rev. Mater. Res.* **2014**, *44*, 241–267. [CrossRef]

10. Huang, H.; Ge, X.; Yu, X.; Jiang, Y.; Peng, Q. Atomistic Studies of Helium Trapping and Diffusion at Ni–Graphene Interfaces. *Eur. Phys. J. Plus* **2025**, *140*, 497. [CrossRef]
11. Ye, Y.; Ma, L.; Tang, T.; Liu, T.; Chen, F.; Ge, X.; Jiang, Y.; Yu, X.; Huang, H. Characterization of Microstructure and Properties of Ti35 Alloy and Its High-Fluence Hydrogen Irradiation-Induced Surface Exfoliation. *Eur. Phys. J. Plus* **2024**, *139*, 822. [CrossRef]
12. Jian, W.-R.; Xu, S.; Su, Y.; Beyerlein, I.J. Energetically Favorable Dislocation/Nanobubble Bypass Mechanism in Irradiation Conditions. *Acta Mater.* **2022**, *230*, 117849. [CrossRef]
13. Fabritsiev, S.A.; Zinkle, S.J.; Singh, B.N. Evaluation of Copper Alloys for Fusion Reactor Divertor and First Wall Components. *J. Nucl. Mater.* **1996**, *233*, 127–137. [CrossRef]
14. Zhang, J. *Molecular Dynamics Study of Effects of Vacancy on Phonon Heat Conductivity of Copper*; Cranfield University: Cranfield, UK, 2015.
15. Ye, X.B.; He, Z.H.; Pan, B.C. The Thermal Conductivity of Defected Copper at Finite Temperatures. *J. Mater. Sci.* **2020**, *55*, 4453–4463. [CrossRef]
16. Fabritsiev, S.A.; Pokrovsky, A.S. The Effect of Neutron Irradiation on the Electrical Resistivity of High-Strength Copper Alloys. *J. Nucl. Mater.* **1997**, *249*, 239–249. [CrossRef]
17. Fabritsiev, S.A.; Pokrovsky, A.S. Effect of High Doses of Neutron Irradiation on Physico-Mechanical Properties of Copper Alloys for ITER Applications. *Fusion Eng. Des.* **2005**, *73*, 19–34. [CrossRef]
18. Zhou, Y.; Anglin, B.; Strachan, A. Phonon Thermal Conductivity in Nanolaminated Composite Metals via Molecular Dynamics. *J. Chem. Phys.* **2007**, *127*, 184702. [CrossRef]
19. Huang, H.; Tang, X.; Xie, K.; Peng, Q. Enhanced Self-Healing of Irradiation Defects near a Ni–Graphene Interface by Damaged Graphene: Insights from Atomistic Modeling. *J. Phys. Chem. Solids* **2021**, *151*, 109909. [CrossRef]
20. Wang, Q.; Gui, N.; Huang, X.; Yang, X.; Tu, J.; Jiang, S. The Effect of Temperature and Cascade Collision on Thermal Conductivity of 3C-SiC: A Molecular Dynamics Study. *Int. J. Heat Mass Transf.* **2021**, *180*, 121822. [CrossRef]
21. Ma, L.; Liu, T.; Cai, B.; Liu, Z.; Zhang, G.; Li, J.; Li, H.; Huang, H. Molecular Dynamics Studies of Primary Irradiation Damage in A-type Ti35 Alloy. *Phys. Status Solidi B* **2023**, *260*, 2200560. [CrossRef]
22. Si, S.; Wang, J.; Li, J.; Li, W.; Cong, H.; Liu, J.; Tang, J.; Jiang, C.; Xia, R.; Xiao, X. Enhancing Resistance to Radiation Hardening and Radiation Thermal Conductivity Degradation by Tungsten/Graphene Interface Engineering. *J. Nucl. Mater.* **2020**, *539*, 152348. [CrossRef]
23. Zhang, H.; Wang, Y.; Sun, J.; Qin, M.; Stirner, T. Lattice Thermal Conductivity of Defected Tungsten Evaluated by Equilibrium Molecular Dynamics Simulation. *Mater. Today Commun.* **2023**, *36*, 106495. [CrossRef]
24. Petersson, C.L.M.; Fredriksson, A.; Melin, S.; Ahadi, A.; Hansson, P. A Molecular Dynamics Study on the Influence of Vacancies and Interstitial Helium on Mechanical Properties of Tungsten. *J. Nucl. Mater.* **2023**, *580*, 154378. [CrossRef]
25. Sharma, S.S.; Parashar, A. Effect of Helium on Thermal Transport Properties in Single- and Bi-Crystals of Ni: A Study Based on Molecular Dynamics. *J. Phys. D Appl. Phys.* **2024**, *57*, 055304. [CrossRef]
26. Tyler, S.K.; Goodhew, P.J. The Growth of Helium Bubbles in Niobium and Nb-1% Zr. *J. Nucl. Mater.* **1978**, *74*, 27–33. [CrossRef]
27. Wei, Q.; Li, N.; Sun, K.; Wang, L.M. The Shape of Bubbles in He-Implanted Cu and Au. *Scr. Mater.* **2010**, *63*, 430–433. [CrossRef]
28. Li, S.-H.; Li, J.-T.; Han, W.-Z. Radiation-Induced Helium Bubbles in Metals. *Materials* **2019**, *12*, 1036. [CrossRef]
29. Liu, T. Primary Irradiation Damage in Ni–Graphene Nanocomposites with Pre-Existing Hydrogen: Insights from Atomistic Simulations. *Eur. Phys. J. Plus* **2024**, *139*, 22. [CrossRef]
30. Shargh, A.K.; Bosić, O.; Abdolrahim, N. Mechanisms of Helium Nanobubble Growth and Defect Interactions in Irradiated Copper: A Molecular Dynamics Study. *J. Nucl. Mater.* **2023**, *574*, 154199. [CrossRef]
31. Hu, L.; Wirth, B.D.; Maroudas, D. Thermal Conductivity of Tungsten: Effects of Plasma-Related Structural Defects from Molecular-Dynamics Simulations. *Appl. Phys. Lett.* **2017**, *111*, 081902. [CrossRef]
32. Plimpton, S. Fast Parallel Algorithms for Short-Range Molecular Dynamics. *J. Comput. Phys.* **1995**, *117*, 1–19. [CrossRef]
33. Stukowski, A. Visualization and Analysis of Atomistic Simulation Data with OVITO–the Open Visualization Tool. *Model. Simul. Mater. Sci. Eng.* **2010**, *18*, 015012. [CrossRef]
34. Kashinath, A.; Demkowicz, M.J. A Predictive Interatomic Potential for He in Cu and Nb. *Model. Simul. Mater. Sci. Eng.* **2011**, *19*, 035007. [CrossRef]
35. Sæther, S.; Erichsen, M.F.; Xiao, S.; Zhang, Z.; Lervik, A.; He, J. Phonon Thermal Transport in Copper: The Effect of Size, Crystal Orientation, and Grain Boundaries. *AIP Adv.* **2022**, *12*, 065301. [CrossRef]
36. Evteev, A.V.; Momenzadeh, L.; Levchenko, E.V.; Belova, I.V.; Murch, G.E. Molecular Dynamics Prediction of Phonon-Mediated Thermal Conductivity of f.c.c. Cu. *Philos. Mag.* **2014**, *94*, 731–751. [CrossRef]
37. Zhang, H.; Sun, J.; Wang, Y.; Stirner, T.; Hamid, A.Y.; Sang, C. Study of lattice thermal conductivity of tungsten containing bubbles by molecular dynamics simulation. *Fusion Eng. Des.* **2020**, *161*, 112004. [CrossRef]
38. Ding, Y.; Wu, X.; Zhan, J.; Chen, Z.; Mao, S.; Ye, M. Simulation study of effects of grain boundary and helium bubble on lattice thermal resistance of tungsten. *Fusion Eng. Des.* **2021**, *168*, 112682. [CrossRef]

39. Huang, H.; Zhong, Y.; Cai, B.; Wang, J.; Liu, Z.; Peng, Q. Size- and Temperature-Dependent Thermal Transport across a Cu–diamond Interface: Non-Equilibrium Molecular Dynamics Simulations. *Surf. Interfaces* **2023**, *37*, 102736. [CrossRef]
40. Lima, H.S.; Tsallis, C. Ising Chain: Thermal Conductivity and First-Principle Validation of Fourier's Law. *Physical A* **2023**, *628*, 129161. [CrossRef]
41. Zou, H.; Feng, Y.; Qiu, L. Excellent Heat Transfer Enhancement of CNT-Metal Interface by Loading Carbyne and Metal Nanowire into CNT. *Int. J. Heat Mass Transf.* **2022**, *186*, 122533. [CrossRef]
42. Al Hossain, J.; Kim, B. Scale Effects in the Nanoscale Heat Transfer of Molecular Interfaces with Different Lattice Orientations. *AIP Adv.* **2021**, *11*, 125311. [CrossRef]
43. Qiu, L.; Li, F.; Zhu, N.; Feng, Y.; Zhang, X.; Zhang, X. Broad Low-Frequency Phonon Resonance for Increased across-Tube Heat Transport. *Phys. Rev. B* **2022**, *105*, 165406. [CrossRef]
44. Bi, S.; Zhao, T.; Tuo, J.; Wang, X.; Sun, Y.; Zhang, Y. Molecular Dynamics Simulation of the Thermal Conductivity and Dielectric Properties of PI/FG Composites. *Surf. Interfaces* **2025**, *56*, 105569. [CrossRef]
45. Liu, F.; Liu, Y.; Jiang, X.Z.; Xia, J. Exploring Thermophysical Properties of CoCrFeNiCu High Entropy Alloy via Molecular Dynamics Simulations. *Heliyon* **2024**, *10*, e36064. [CrossRef]
46. Chen, N.; Yang, K.; Wang, Z.; Zhong, B.; Wang, J.; Song, J.; Li, Q.; Ni, J.; Sun, F.; Liu, Y.; et al. Quantifying Interfacial Bonding Using Thermal Boundary Conductance at Cubic Boron Nitride/Copper Interfaces with a Large Mismatch of Phonon Density of States. *ACS Appl. Mater. Inter.* **2023**, *15*, 34132–34144. [CrossRef] [PubMed]
47. Hu, Y.; Xu, J.; Ruan, X.; Bao, H. Defect Scattering Can Lead to Enhanced Phonon Transport at Nanoscale. *Nat. Commun.* **2024**, *15*, 3304. [CrossRef] [PubMed]
48. Saxena, V.K.; Saxena, S.C. Measurement of the Thermal Conductivity of Helium Using a Hot-Wire Type of Thermal Diffusion Column. *J. Phys. D Appl. Phys.* **1968**, *1*, 1341–1351. [CrossRef]
49. Lee, C.-W.; Chernatynskiy, A.; Shukla, P.; Stoller, R.E.; Sinnott, S.B.; Phillpot, S.R. Effect of Pores and He Bubbles on the Thermal Transport Properties of UO_2 by Molecular Dynamics Simulation. *J. Nucl. Mater.* **2015**, *456*, 253–259. [CrossRef]
50. Zhang, C.W.; Zhou, H.; Zeng, Y.; Zheng, L.; Zhan, Y.L.; Bi, K.D. A Reduction of Thermal Conductivity of Non-Periodic Si/Ge Superlattice Nanowire: Molecular Dynamics Simulation. *Int. J. Heat Mass Transf.* **2019**, *132*, 681–688. [CrossRef]
51. Estreicher, S.K.; Gibbons, T.M.; Kang, B.; Bebek, M.B. Phonons and Defects in Semiconductors and Nanostructures: Phonon Trapping, Phonon Scattering, and Heat Flow at Heterojunctions. *J. Appl. Phys.* **2014**, *115*, 012012. [CrossRef]

materials

MDPI

Article

Machine Learning-Driven Design and Optimization of Multi-Metal Nitride Hard Coatings via Multi-Arc Ion Plating Using Genetic Algorithm and Support Vector Regression

Yu Gu [1,2,*], Jiayue Wang [3], Jun Zhang [1], Yu Zhang [1,2], Bushi Dai [1,2], Yu Li [4], Guangchao Liu [1], Li Bao [5] and Rihuan Lu [6]

1 College of Mechanical Engineering, Shenyang University, Shenyang 110044, China; tongxun_e@126.com (J.Z.); 670003099@163.com (Y.Z.); bushi0509@126.com (B.D.); liugclxy@126.com (G.L.)
2 Laboratory of Research and Application of Multiple Hard Films, Shenyang University, Shenyang 110044, China
3 Informatization Office, Shenyang University, Shenyang 110044, China; jiayue.wang@syu.edu.cn
4 College of Intelligence and Informatics, Shenyang University, Shenyang 110044, China; w179yxl@syu.edu.cn
5 School of Mechanical and Electrical Engineering, Qiqihar University, Qiqihar 161000, China; lwjxs2019@163.com
6 National Engineering Research Center for Equipment and Technology of Cold Rolled Strip, Yanshan University, Qinhuangdao 066004, China; lrh@ysu.edu.cn
* Correspondence: first_e@126.com

Abstract: The goal of this study is to develop an efficient machine learning framework for designing high-hardness multi-metal nitride coatings, overcoming the limitations of traditional trial-and-error methods. The development of multicomponent metal nitride hard coatings via multi-arc ion plating remains a significant challenge due to the vast compositional search space. Although theoretical studies in macroscopic, mesoscopic, and microscopic domains exist, these often focus on idealized models and lack effective coupling across scales, leading to time-consuming and labor-intensive traditional methods. With advancements in materials genomics and data mining, machine learning has become a powerful tool in material discovery. In this work, we construct a compositional search space for multicomponent nitrides based on electronic configuration, valence electron count, electronegativity, and oxidation states of metal elements in unary nitrides. The search space is further constrained by FCC crystal structure and hardness theory. By incorporating a feature library with micro-, meso-, and macro-structural characteristics and using clustering analysis with theoretical intermediate variables, the model enriches dataset information and enhances predictive accuracy by reducing experimental errors. This model is successfully applied to design multicomponent metal nitride coatings using a literature-derived database of 233 entries. Experimental validation confirms the model's predictions, and clustering is used to minimize experimental and data errors, yielding a strong agreement between predicted optimal molar ratios of metal elements and nitrogen and measured hardness performance. Of the 100 Vickers hardness (HV) predictions made by the model using input features like molar ratios of metal elements (e.g., Ti, Al, Cr, Zr) and atomic size mismatch, 82 exceeded the dataset's maximum hardness, with the best sample achieving a prediction accuracy of 91.6% validated against experimental measurements. This approach offers a robust strategy for designing high-performance coatings with optimized hardness.

Keywords: machine learning; small sample problem; multi-arc ion plating; XN coating; hardness

1. Introduction

Multi-metal nitride hard coatings are widely used in industrial applications like cutting tools, molds, and high-performance surface engineering due to their excellent wear, corrosion resistance, and high-temperature stability [1–7]. They can withstand extreme conditions, being crucial in environments with mechanical stress, high temperatures, and corrosive elements. However, designing these coatings for optimal performance is challenging. This is mainly because of the complexity of their composition systems and intricate microstructures [8–10].

Traditional material development mainly relies on trial-and-error experiments, which are time- and resource-consuming, involving repeated synthesis and testing. Guided by macro-, meso-, and micro-level theories, these approaches lack integration among theoretical frameworks [11,12]. The gap between theoretical predictions and experimental validation causes inefficiencies in development, slowing down new material discovery and optimization. In multi-arc ion plating for multi-metal nitride coatings, the challenge is greater due to the vast compositional search space. Considering multiple metallic elements and process parameters like temperature, pressure, and deposition time, the system complexity grows exponentially, making it hard to explore all potential configurations [13,14].

To tackle these challenges, researchers are exploring computational approaches to complement traditional experiments. With the rising significance of computational materials science, especially materials informatics and data mining, material design has entered a new era. These techniques allow for more efficient exploration of material composition and process spaces, reducing the dependence on pure experimentation [15–18]. Recent computational method advances have transformed material discovery and optimization. A key development area is applying machine learning (ML) and algorithms like genetic algorithms (GAs) to materials science. ML models using large datasets can predict material properties from input parameters such as composition and processing conditions [19–26]. Using these models, researchers can predict new material behavior before synthesis, saving time and resources.

For example, genetic algorithms (GAs) are widely used in material optimization, especially for finding the best combination of compositional and process variables. GAs mimic natural selection, using crossover, mutation, and selection to solve optimization problems over time [19,20]. They are effective in materials design as they can explore numerous potential solutions quickly. Combined with physical models at the atomic or molecular level, GAs optimize coating composition and heat treatment conditions [21,22]. Moreover, support vector regression (SVR), an ML model based on statistical learning theory, predicts material properties accurately. SVR is suitable for predicting continuous outcomes like hardness or wear resistance of multi-metal nitride coatings [23–28]. Its strength is finding the optimal hyperplane for data fitting, making it great for property prediction in complex systems.

Combining GAs and SVR enables researchers to predict the performance of ternary metal nitrides in multi-metal systems. This combined method allows for rapid screening of potential material configurations, speeding up discovery. The GAs with an elite strategy retain top-performing solutions during optimization, and the SVR model improves property prediction accuracy by using intermediate variables from physical theories, refining data for better exploration of the compositional space.

Nonetheless, challenges and controversies exist in multi-metal nitride hard coating design. A key problem is the dataset size and quality for ML models. ML models like SVR perform best with large and high-quality data. However, in multi-metal nitride design, data are often scarce due to experiment costs, time, or data privacy. Small dataset models may overfit and struggle to generalize.

Moreover, integrating physical models into ML frameworks shows potential but is debated. Incorporating theoretical insights as intermediate variables is a controversial challenge. Some researchers argue that existing theories, such as those describing the relationships between composition, microstructure, and properties, are not fully utilized in current ML frameworks [25,26]. They contend that by more deeply embedding these theories into ML models, the predictions could be made more reliable and interpretable. However, doing so introduces additional complexity, making the models harder to train and increasing the computational resources required.

Another controversial area is the variability of experimental conditions. The performance of multi-metal nitride coatings is very sensitive to synthesis and testing conditions like temperature, pressure, and deposition time. Small parameter variations can cause significant property differences, making modeling and prediction difficult. Some researchers doubt if current ML models can capture this variability and wonder if new approaches are needed to consider the complex interactions between composition and processing conditions. To tackle these challenges, our study introduces a GA-SVR methodology that enhances prediction accuracy by integrating physical models and addressing data limitations. To address challenges of small datasets, integrating physical models with ML, and experimental condition variability, the GA-SVR methodology in this study is a significant advancement in multi-metal nitride coating design. By using existing theories as intermediate variables, it enhances data quality and prediction accuracy, enabling more efficient exploration of complex compositional systems, which is crucial in multi-metal systems where traditional experiments are time-consuming and costly.

This approach not only shows potential for better prediction accuracy but also can shorten the development cycle. In industry, quickly identifying promising candidates and validating performance is a major plus. The methodology offers a valuable framework for future R&D in material design.

This study begins by establishing a search space based on the micro- and mesoscopic properties of unary nitride hardness (Section 2.1), followed by feature engineering using solid solution strengthening theory (Section 2.2) and the development of the GA-SVR model (Section 2.4). Building on this foundation, the prediction model is constructed (Section 3), starting with the GA's fitness function, chromosome coding for alloy compositions (e.g., molar ratios of metal elements), and population initialization (Section 3.1), linking computational optimization to alloy design. Section 4 (Results) presents the model's predictions, and Section 5 (Prediction Results and Experimental Validation) validates these predictions against experimental hardness measurements.

In conclusion, this study aimed to develop an efficient and accurate machine learning framework combining a genetic algorithm (GA) and support vector regression (SVR) to design high-hardness multi-metal nitride coatings, addressing the inefficiencies of traditional trial-and-error methods. It has advantages over traditional methods, reducing discovery time and cost and enabling more accurate property predictions. The study's insights benefit the multi-metal nitride coating field and offer a broader basis for future material design innovations.

The need for additional research arose from the limitations of traditional trial-and-error methods, which are inefficient and costly for designing multi-metal nitride coatings due to the vast compositional search space and sensitivity to experimental conditions such as temperature and pressure. Prior studies often utilized idealized models that failed to integrate micro-, meso-, and macro-scale properties or account for experimental variability, resulting in suboptimal predictive accuracy. Furthermore, the small dataset size (233 entries, Section 2.1) compared to large datasets (tens of thousands to millions) or medium datasets (hundreds to thousands) in other machine learning applications posed a unique challenge. This study addresses these gaps by developing a GA-SVR framework that leverages advanced feature engineering and optimization to enhance predictive accuracy and efficiency, enabling rapid exploration of complex multicomponent nitride systems.

2. Problem Description of Hardness Prediction Model

2.1. Establishment of Searching Space

To build the search space for metal elements, this study identified high-hardness unary metal nitrides via a comprehensive literature survey. Figure 1 illustrates a high-hardness unary nitride (e.g., TiN), highlighting its key properties such as valence electron count and electronegativity, which guide the selection of metal elements (e.g., Ti, Al, Cr, Zr) for constructing the multicomponent nitride search space in the GA-SVR model. The selected metal elements form the set ME (Equation (1)), which is the search space for nitrides in this study. In the nitrides system, the micro-mechanisms of unary nitride hardness, including chemical bonding and electronic structures, provide a fundamental theoretical basis for understanding multicomponent nitride performance [29–34]. These mechanisms offer essential insights for analyzing the complex interactions determining multicomponent nitride properties.

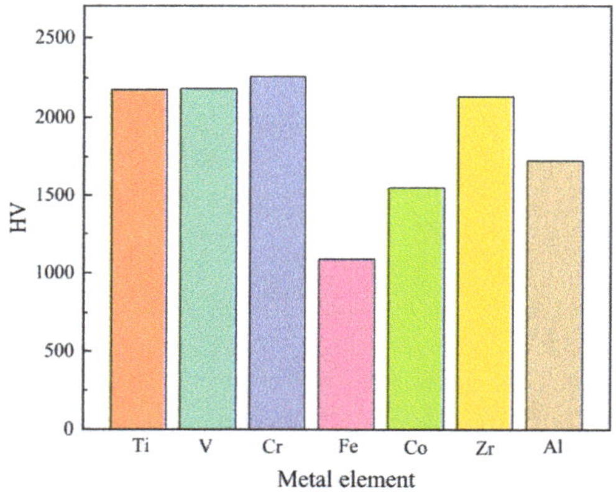

Figure 1. Hardness of single metal nitrides.

An analysis of the seven metal elements in ME reveals shared characteristics, including outermost electron orbitals of 4s, 3d, or 4d; electronegativity values ranging from 1.54 to 1.91; and common oxidation states of +1, +2, +3, +4, +5, and +6. These similarities in electronic configuration, valence electron count, electronegativity, and oxidation states significantly influence the physical and chemical properties of metal nitrides, such as hardness, stability, and reactivity [32,33]. Based on these findings, the set MP (Equation (2)), comprising these four attributes, is incorporated into the model's search space and utilized as critical features in the model.

$$ME = \{\mathrm{Ti}, \mathrm{V}, \mathrm{Cr}, \mathrm{Fe}, \mathrm{Co}, \mathrm{Zr}, \mathrm{Al}\} \tag{1}$$

$$MP = \{(E_i, V_i, X_i, O_i) | i \in M\} \tag{2}$$

Here, E_i represents the electronic configuration, V_i the valence electron count, X_i the electronegativity, and O_i the oxidation state of each metal element in M. The mapping function f establishes the relationship from MP to ME (Equation (3)).

$$f : ME \rightarrow MP \tag{3}$$

This study focuses on the strengthening mechanisms of multicomponent FCC-structured nitrides under specific experimental conditions and datasets [35,36]. A subset of *ME*, denoted as ME_Sub (Equation (4)), is derived to define the search space for metal elements in the model, with its relationship to *ME* expressed as Equation (5).

$$ME\_Sub = \{\text{Cr, Ti, Al, Zr}\} \tag{4}$$

$$ME\_Sub \subseteq ME \tag{5}$$

To support the research, a dataset was compiled from the literature, encompassing FCC metal nitrides prepared via the multi-arc ion plating method. This dataset contains 233 entries, including 4 unary metal nitrides, 34 binary metal nitrides, and 195 ternary metal nitrides. These data provide a robust experimental foundation and theoretical framework for the performance prediction and optimization of multicomponent nitrides.

2.2. Problem Description and Model Assumptions

According to the "No Free Lunch Theorem", it is challenging to identify an optimal ML model applicable to all materials problems [37]. To enhance the predictive accuracy of the model, we constructed a feature space based on solid solution strengthening theory [38,39]. Thus, incorporating the solid solution strengthening mechanism into the SVR model, along with relevant constraints added through the GA as the top-level optimization algorithm, is critical for improving predictive performance.

To establish these constraints, the following derivations are made. Assuming the metal nitride is an ideal solid solution and all crystal structures in the selected dataset are FCC, the predictions are based on this FCC structure. The derivation proceeds as follows:

Case 1: Ternary Metal Nitride as Ideal Substitutional Solid Solutions

In an ideal substitutional solid solution, the solute atoms replace solvent atoms in a 1:1 ratio, leading to a constrained relationship between metal atoms and N atoms [40]. Based on this assumption, four states with varying solute concentrations are discussed:

Low Solute Concentration: On the condition of low solute concentrations, solute atoms randomly substitute solvent atoms, forming a disordered solid solution. In this state, solvent atoms still dominate the lattice structure, and the ratio of solute to solvent atoms does not induce significant lattice changes [41].

Moderate Solute Concentration: As the solute concentration increases, solute atoms start to significantly affect lattice properties, such as lattice constants, electronic structure, and material properties. Although the contribution of solute atoms increases, the lattice structure remains primarily controlled by the solvent atoms, and solute atoms have not yet altered the dominant characteristics of the lattice [39,40]. The relationship between metal atoms and N atoms at low and moderate solute concentrations can be expressed as Equation (6):

$$\frac{(a_1 + b_1)}{c_1} = \frac{(a_2 + b_2)}{c_2} \tag{6}$$

where a_n represents the moles of solvent atoms, b_n represents the moles of solute atoms, and c_n represents the moles of N atoms, with n denoting the current state.

Exceeding Solid Solution Limit: When the solute concentration reaches the critical point (the solid solution limit), the lattice structure undergoes significant changes. At this stage, the interaction between solute and solvent atoms becomes more balanced, with the solute atoms sometimes dominating certain physical properties, forming a solute–solvent synergistic effect [41,42]. When the solute concentration exceeds this limit, the lattice structure changes, and solute atoms take over the lattice.

High Solute Concentration: At very high solute concentrations, solute atoms dominate the lattice structure, surpassing the solvent atoms, which are now effectively "dilute solutes". This typically occurs in alloy systems where the solute and solvent have highly similar physical and chemical properties [41].

In the third and fourth cases, the interaction between metal and N atoms shifts from weak to stronger associations, signifying a structural change in the positions and roles of solute and solvent atoms in the lattice. In this context, the relationship between metal atoms and N atoms is expressed as Equation (7):

$$\frac{(a_3+b_3)}{c_3} = \frac{(a_4+b_4)}{c_4} \neq \frac{(a_1+b_1)}{c_1} \tag{7}$$

Case 2: Ternary Metal Nitride as Complete Interstitial Solid Solutions

In contrast to substitutional solid solutions, in complete interstitial solid solutions, the solute atoms occupy interstitial positions in the solvent crystal lattice, rather than directly replacing the solvent atoms. Due to their smaller size, interstitial atoms can be accommodated in the lattice voids, giving these solutions a solubility limit [41]. Based on this characteristic, the relationship between metal atoms and N atoms in ternary metal nitride is expressed as Equation (8):

$$\frac{a}{c} < \frac{(a+b)}{c} \leq \frac{(1+\delta_{\max})a}{c} \tag{8}$$

where a, b, and c represent the moles of solvent atoms, solute atoms, and N atoms, respectively, and $\delta_{\max}$ is a constant taken as 0.022 based on the solubility limits of typical interstitial solid solutions [43,44].

Case 3: Mixed Solid Solutions

In mixed solid solutions, both substitutional and interstitial solid solutions coexist. In this case, the relationship between metal atoms and N atoms in ternary metal nitride is similar to that in Case 2 and is expressed by Equation (8), as no further distinction is necessary.

2.3. The Establishment of Model Constraints

The proposed model, based on the search space defined in Section 2.1, aims to predict the hardness of ternary metal nitrides. As analyzed in Section 2.2, the atomic composition of ternary metal nitrides is closely related to the corresponding unary nitrides. Therefore, we first selected unary nitrides with FCC lattices for Ti, Al, Cr, and Zr from the Materials Project (https://materialsproject.org/, accessed on 4 June 2025), forming a subset labeled "subset-N" (see Supplementary Table S1 for details).

Building on the results from Section 2.2, the model's constraints are defined. The molar ratios of the elements are used to represent their proportions in the compound, denoted as x_1, x_2, x_3, x_4, and x_5 for Ti, Al, Cr, Zr, and N, respectively. To ensure the formation of nitrides, the molar ratio of N element (x_5) must be nonzero. Following the above analysis 1, we formulate Constraint 1 and Constraint 2, which are expressed as follows:

$$\text{Constraint 1}: \sum_{i=1}^{n} x_i = 1 \ \ (\text{if and only if } x_i = 0,\ (i = 1,2,3,4) \tag{9}$$

$$\text{Constraint 2}: \eta_{ratio} < \frac{\sum_{i=1}^{4} x_i}{x_5} \leq \eta_{ratio} + \delta_{\max} \quad (\eta_{ratio} \in subset - N) \tag{10}$$

2.4. Selection of Model Features and Establishment of Objective Function

Unlike traditional ML methods, which typically rely on raw inputs such as composition and process parameters to directly associate with target output properties [45,46], this study introduces a feature library that systematically incorporates micro-, meso-, and macro-structural characteristics of materials. These structural features play a crucial role in bridging the raw inputs and target outputs and must be considered comprehensively [47–49].

In principle, the more comprehensive the features included in the model, the more complete the search space, which in turn improves prediction accuracy and coverage [50]. Therefore, this study broadens the parameter space by collecting as many features as possible and converts these features into descriptors recognizable by ML algorithms, enhancing the model's predictive capability. Based on elemental parameters, we developed two statistical models (Equations (11) and (12)) to calculate statistical data for multicomponent compounds. For subsequent analysis, all features were grouped into four subsets (Supplementary Table S2).

$$\overline{x} = \sum c_i x_i \quad (11)$$

$$\delta_x = \sqrt{\sum c_i (1 - x_i/\overline{x})} \quad (12)$$

where c_i represents the molar percentage of atoms and x_i is the characteristic of component I; if $\overline{x}$ is zero, then $x_i/\overline{x}$ is also zero.

Once the dataset and feature sets are constructed, an SVR model can be established. SVR, a regression method based on statistical learning theory, fits data by constructing a hyperplane with a maximum margin, ensuring that the distance from all training samples to the hyperplane is within a given ε [0.05, 0.2] range, expressed as follows (Equation (13)).

$$K(x_i, x_j) = \exp\left(-\frac{|x_i - x_j|^2}{2\sigma^2}\right) \quad (13)$$

where σ is the kernel parameter that controls the width of the Gaussian distribution. The normalization condition is applied to the input features x_i and x_j, which are scaled to [0,1] to ensure consistent kernel behavior across differing feature scales (e.g., molar ratios, valence electron count). However, $K(x_i, x_j)$ itself is not normalized, as its output is inherently bounded by the RBF kernel and controlled by σ, avoiding redundancy and preserving the intended mapping.

In most ML models, the objective function is typically a clear mathematical expression designed to directly optimize a specific performance metric. However, in this study, the objective function employs the SVR method [51,52]. This design contrasts with traditional explicit objective functions, as SVR establishes a data-driven nonlinear regression model to achieve optimization. By using SVR as the objective function, the model performs nonlinear mapping in a high-dimensional feature space, allowing it to more flexibly adapt to the complex relationship between input features and target outputs [53]. The advantage of this approach is that SVR not only captures the nonlinear characteristics of the data but also enhances the model's generalization ability by maximizing the margin, thus achieving an effective balance between prediction accuracy and model robustness.

3. The Establishment of an Intensity Prediction Model Algorithm

The model presented in this study integrates two algorithms: GA with an elitist strategy and SVR. The GA, with its robust global search capability, effectively avoids local optima in complex solution spaces, offering broad exploration during the optimization process. In contrast, SVR, a regression technique that partitions data to search for a hyperplane, features a simple structure and demonstrates high computational efficiency, particularly on small datasets, allowing for rapid model training and prediction. The combination of these two algorithms leverages the GA's global search advantages and the SVR's efficiency with small datasets, thereby enhancing both the prediction accuracy and computational efficiency of the model. Figure 2 illustrates the architecture of the GA-SVR model, where the GA (top) optimizes the feature set, including molar ratios of Ti, Al, Cr, Zr, and derived structural characteristics (e.g., electronic configuration, valence electron count), and passes these to the SVR (bottom) for hardness prediction. The figure also shows the training process, where the dataset of 233 entries is used to train the SVR model, with the GA iteratively refining inputs to minimize prediction errors.

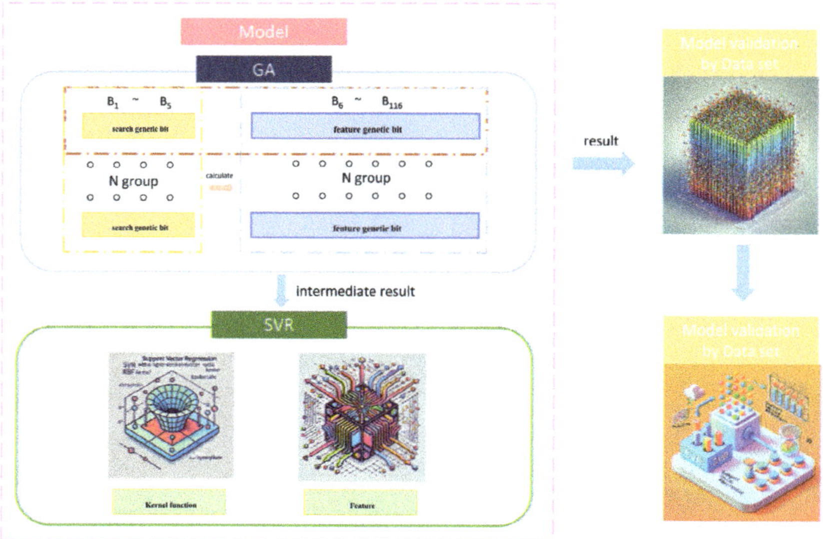

Figure 2. Schematic diagram of hardness prediction model algorithm.

3.1. Fitness Function of Strength Prediction Model, Chromosome Coding, and Population Initialization

This study presents a model built upon GA and SVR. A fitness function is constructed to assess the quality of candidate solutions, with the primary objective of predicting the hardness of ternary metal nitrides. In this model, the output of the objective function represents the required hardness prediction, thereby linking the objective function with the fitness function and ensuring functional consistency, as exemplified in the SVR model developed in Section 2.3.

The direct relationship between the objective and fitness functions ensures that the genetic algorithm effectively explores the solution space. Through iterative optimization, it progressively enhances the quality of candidate solutions, thereby achieving the set optimization goal [51,52]. This design not only streamlines the algorithm's implementation, focusing on the core problem of hardness prediction, but also enhances the algorithm's transparency and interpretability, providing a solid theoretical foundation for the reliability of the research outcomes.

As discussed in Section 2.2, the type of solid solution influences the associated constraints, necessitating the identification of the solid solution type before encoding. The formation of solid solutions is governed by five key factors: atomic size similarity, chemical affinity, valence electron concentration, lattice type compatibility, and thermodynamic conditions. In this study, atomic size similarity is used as the primary criterion for determining the solid solution type, as described in Equation (14):

$$|r_{solute} - r_{solvent}| \leq \Delta r_{max} \tag{14}$$

Here, r_{solute} represents the atomic radius of the solute; $r_{solvent}$ represents the atomic radius of the solvent; and Δr_{max} represents the maximum permissible atomic radius difference, typically set to 0.15.

Based on the atomic size similarity criterion, a solid solution determination coefficient is constructed as follows:

$$\delta_{ss} = \begin{cases} 0 & all \ \left| r_i - r_j \right| \leq \Delta r_{\max} \quad i,j \in M \\ \delta_{\max} & else \end{cases} \tag{15}$$

where M represents the set of atomic radii for the metal elements involved in the model, specifically Cr, Ti, Al, and Zr.

$$\begin{aligned} & f(\mathrm{B}_{k1}) = 0 \\ & f(\eta_{ration}, \delta_{ss}) = \frac{\eta_{ration}}{1+\eta_{ration}} + \theta\left(\frac{\eta_{ration}+\delta_{ss}}{1+\eta_{ration}+\delta_{ss}} - \frac{\eta_{ration}}{1+\eta_{ration}}\right) \\ & f(\mathrm{B}_{k2}) = \lambda_1 f(\eta_{ration}, \delta_{ss}) \\ & f(\mathrm{B}_{k3}) = \lambda_2 f(\eta_{ration}, \delta_{ss}) \\ & f(\mathrm{B}_{k4}) = \lambda_3 f(\eta_{ration}, \delta_{ss}) \\ & f(\mathrm{B}_5) = 1 - f(\eta_{ration}, \delta_{ss}) \\ & k_i \in \{1,2,3,4\} \quad k_i \neq k_j \quad i,j = 1,2,3,4 \quad \lambda_1 + \lambda_2 + \lambda_3 = 1 \qquad 0 < \theta \leq 1 \end{aligned} \tag{16}$$

In general, population initialization in GA is achieved by random generation of individuals. However, due to the need to satisfy constraints 1 and 2 in the model, and to minimize the waste code rate, specific initialization methods are applied to the first five positions of the chromosome. Specifically, these positions are initialized based on Equation (16), ensuring that the genes of each individual conform to the solid solution type criteria.

The hardness calculation model consists of two main components (Figure 3); the first involves searching within the molar ratio space of Ti, Al, Cr, Zr, and N elements in the ternary metal nitride system, and the second involves computing the features of the identified compounds to evaluate their fitness. Therefore, the chromosome structure is divided into two parts. The first part, from the first to the fifth gene positions, corresponds to the molar ratios of the aforementioned elements, termed the "search gene positions". The second part, from the 6th to the 45th gene positions, calculates the relevant features of the nitrides identified in the search, using Equation (11), while the 46th to the 85th positions are derived using Equation (12). Notably, all gene positions in the chromosome are encoded in decimal format.

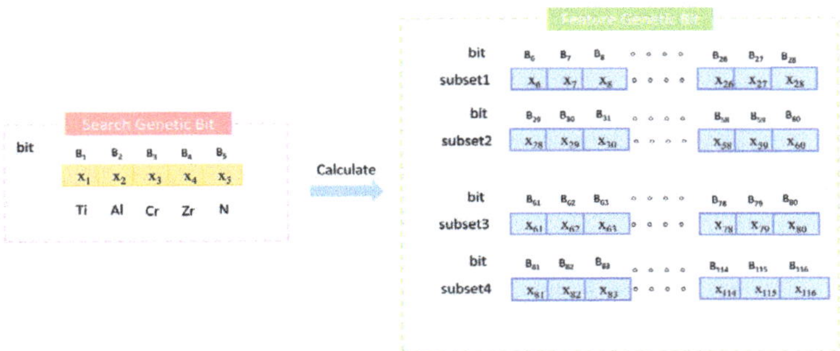

Figure 3. Schematic diagram of population initialization in algorithm.

In summary, the population initialization method described above generates the chromosomes for each individual in the population. After initialization, each chromosome represents an initial solution to the model, forming a population of 100 chromosomes that serves as the foundation for the subsequent genetic algorithm optimization process.

3.2. Chromosome Roulette Selection and Crossover Operator Design

The selection of chromosomes within the population is a crucial step in the evolutionary process of GA. In this model, a roulette wheel selection strategy is employed to select the chromosomes for the next generation according to Equation (17).

$$P_i = \frac{F_i}{\sum\limits_{i=1}^{N} F_i} \quad i \in N \tag{17}$$

$$P_i^{sum} = \sum_{k=i}^{i} P_k \quad i \in N \tag{18}$$

Traditional crossover operators, such as single-point or uniform crossover, are not directly applicable in this model due to the unique requirements of decimal encoding and the specific stoichiometric and physical constraints of multi-metal nitride compositions. Therefore, an innovative crossover algorithm suitable for decimal encoding and model constraints is proposed (Equation (19)). This algorithm combines the advantages of multi-point and arithmetic crossover, leveraging the flexibility of multi-point crossover to select multiple gene segments and the precision of arithmetic crossover to blend numerical values, thus preserving the continuity of decimal-encoded molar ratios. The algorithm is implemented in two stages: First, a genetic crossover stage, where the crossover is carried out based on a five-bit crossover mask. The genes corresponding to the mask value of 1 are crossed using the arithmetic crossover formula. Second, a gene repair stage (Equation (20)) ensures that the newly generated genes conform to the model's constraints, such as maintaining the sum of molar ratios to 1 and ensuring non-negative values, by normalizing or adjusting the genes post-crossover (Section 3.1). This repair mechanism is critical to producing valid offspring that align with the physical and chemical properties of nitride coatings (Figure 4).

$$\begin{aligned} f(C^1{}_{B_i}) &= (f(P^1{}_{B_i}) + \lambda(f(P^1{}_{B_i}) - f(P^2{}_{B_i})) \\ f(C^2{}_{B_i}) &= (f(P^2{}_{B_i}) + \lambda(f(P^2{}_{B_i}) - f(P^1{}_{B_i})) \end{aligned} \tag{19}$$

$$\begin{aligned}
&\eta_{\mathrm{ratio}-C^m} = \eta_{\mathrm{ratio}-p^m} \qquad m = 1,2 \\
&f(\mathrm{B}_{k1-C^m}) = 0 \\
&f(\eta_{ration-C^m}, \delta_{ss}) = \tfrac{\eta_{ration-C^m}}{1+\eta_{ration-C^m}} + \theta'\left(\tfrac{\eta_{ration-C^m}+\delta_{ss}}{1+\eta_{ration-C^m}+\delta_{ss}} - \tfrac{\eta_{ration-C^m}}{1+\eta_{ration-C^m}}\right) \\
&f(\mathrm{B}_{k2-C^m}) = \lambda'_1 f(\eta_{ration-C^m}, \delta_{ss}) \\
&f(\mathrm{B}_{k3-C^m}) = \lambda'_2 f(\eta_{ration-C^m}, \delta_{ss}) \\
&f(\mathrm{B}_{k4-C^m}) = \lambda'_3 f(\eta_{ration-C^m}, \delta_{ss}) \\
&f(\mathrm{B}_{5-C^m}) = f(\mathrm{B}_{5-p^m}) \\
&k_{i-C^m} \in \{1,2,3,4\} \quad k_{i-C^m} \neq k_{j-C^m} \quad i,j = 1,2,3,4 \quad \lambda'_1 + \lambda'_2 + \lambda'_3 = 1 \qquad 0 < \theta' \leq 1
\end{aligned} \tag{20}$$

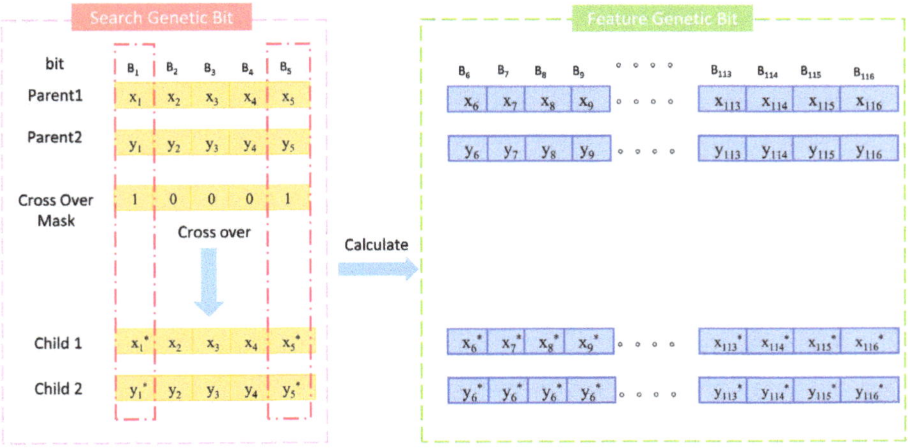

Figure 4. Schematic diagram of crossover strategy in algorithm. A crossover mask value of 1 produces new encodings through crossover, while a value of 0 preserves the original encoding. Thus, the asterisk (*) symbol marks positions where new encodings are generated.

3.3. Variation Operator Design and Elite Strategy

To enhance the stability and robustness of the algorithm, a mutation mechanism is incorporated, comprising two components (Figure 5): the design of a mutation mask and the definition of a mutation probability. Each gene position on the chromosome is assigned an independent mutation probability (Equation (21)). Following mutation, all chromosomes are evaluated against Constraint 1 and Constraint 2 to ensure validity. Any chromosome failing to meet these constraints is repaired using Equation (20).

$$\begin{aligned} & f(B^{*}{}_{B_i}) = f(B_{B_i}) + \Delta f(B_{B_i}) = 0 \quad (i = 1, 2, \ldots\ldots, n) \\ & \Delta f(B_{B_i}) = \theta f(B_{B_i})(-0.3 \leq \theta \leq 0.3) \end{aligned} \tag{21}$$

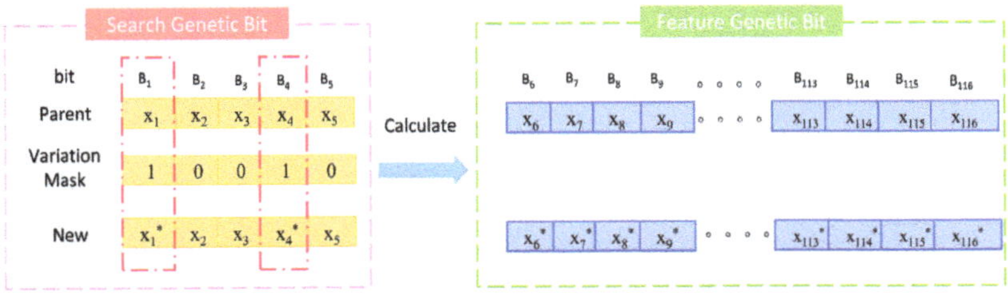

Figure 5. Schematic diagram of variation strategy in algorithm. A mutation mask value of 1 introduces new encodings through mutation, while a value of 0 preserves the original encoding. The asterisk (*) explicitly marks positions where mutations occur.

In the GA, interdependencies among gene positions within an individual are common. Simple and independent application of crossover and mutation operations to each gene position can disrupt beneficial gene combinations, hindering the accumulation of high-quality genetic information. To address this issue, an elite strategy is incorporated, which preserves the genetic information of the best-performing individuals to enhance convergence and global optimization.

The implementation of the elite strategy involves the following steps (Figure 6):

1. Fitness Evaluation: Evaluate the fitness of all individuals in the population and identify the highest-performing individuals.

2. Elite Retention: Directly copy the top 10 individuals with the highest fitness scores to the next generation.
3. Standard Genetic Operations: Apply conventional genetic operations, such as crossover and mutation, to the remaining individuals to generate new candidate solutions.
4. Generation of the New Population: Combine the retained elite individuals with the newly generated candidates to form the next generation.

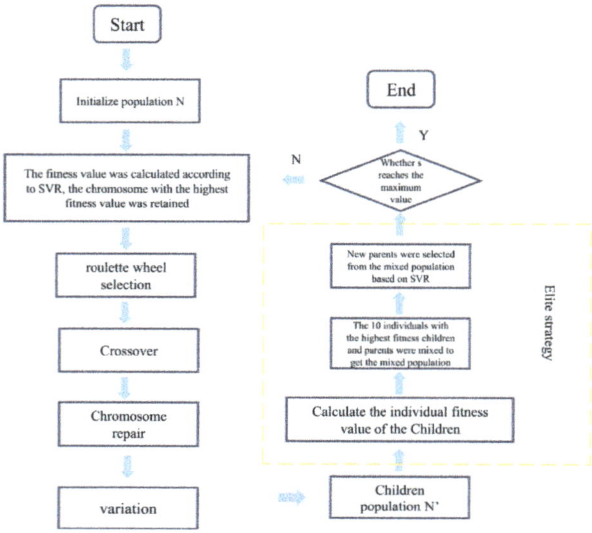

Figure 6. Diagram of the Elite Strategy.

Additionally, a clear stopping criterion is established for the genetic algorithm: the search terminates when the number of iterations reaches 800 generations, at which point the best solution in the current population is output as the final result. By preserving optimal solutions, the elite strategy accelerates convergence and improves global optimization performance, ensuring the algorithm remains robust and efficient when addressing complex optimization problems.

4. Experimental Methods

4.1. Hardness Precipitate Details

This study compiles compositional and hardness data for various ternary metal nitrides reported in the literature, all of which exhibit a face-centered cubic (FCC) crystal structure. To construct the predictive model, the elemental ratios of compounds in a selected dataset were used as the initial input variables. For model optimization, the algorithm parameters were set as follows: a population size of 400, a crossover rate of 0.85, a mutation rate of 0.0035, and 800 iterations. To comprehensively evaluate the model's performance, additional independent datasets were employed for validation. Multiple predictions were performed for each sample to assess the model's generalization capability, ensuring robustness and applicability across different datasets.

4.2. SVR Model Accuracy Detection

To evaluate the generalization ability of the model, the collected dataset was processed as follows. Given the limited size of the dataset (only 233 samples), the way the training and testing sets are split has a significant impact on model performance. Additionally, random data splitting can lead to imbalanced distributions, introducing potential biases in the evaluation. To provide a more comprehensive assessment, this study employed k-fold cross-validation. Specifically, the 233 samples were randomly divided into 10 groups, each containing approximately 70 samples, ensuring that each sample participated in both training and testing multiple times. In each iteration, 9 groups were randomly selected as the training set, while the remaining 1 group served as the test set. This process was repeated 10 times, ensuring that each group was used as a test set at least once.

$$MAE = \frac{1}{n}\sum(x_i - y_i) \tag{22}$$

$$MSE = \frac{1}{n}\sum(x_i - y_i)^2 \tag{23}$$

where n represents the number of samples, and x_i and y_i denote the predicted and actual values for the ith sample, respectively.

$$SST = \sum\left(y_i - \frac{1}{n}\sum y_i\right)^2 \tag{24}$$

$$SSR = \sum\left(x_i - \frac{1}{n}\sum y_i\right)^2 \tag{25}$$

$$R^2 = \frac{SSR}{SST} \tag{26}$$

To quantify the model's generalization capability, we evaluated its performance using metrics including Mean Absolute Error (MAE) and Mean Squared Error (MSE), with calculation formulas defined in Equations (22) and (23). Additionally, the coefficient of determination (R^2) was calculated using the regression sum of squares (SST) and total sum of squares (SSR) to measure the model's ability to explain the variance in the data. The relevant formulas are provided in Equations (24)–(26). Here, SST represents the variance explained by the model, and SSR represents the total variance in the data. The value of R^2 ranges from 0 to 1; the closer the value is to 1, the better the model's explanatory power and fit to the data. Conversely, an R^2 value near 0 indicates a weak explanatory power and poor model fit.

4.3. Materials and Multi-Arc Ion Experiment

M2 high-speed steel substrate (Φ40 mm × 5 mm, pre-treated, Huangshi) was purchased from Daye Special Steel Co., Ltd., Huangshi, Hubei, China. High-purity argon gas (Ar, 99.999%) and nitrogen gas (N_2, 99.999%) were obtained from Shanghai Chinllenge Gases Co., Ltd., Shanghai, China, and Wuhan Newradar Special Gas Co., Ltd., Wuhan, Hubei, China, respectively. Chromium-Titanium (Cr-Ti) alloy target material (99.99% purity) and Aluminum-Zirconium (Al-Zr) alloy target material (99.99% purity) for plasma coating were sourced from Beijing Zhongnuo Advanced Material Technology Co., Ltd., Beijing, China.

The pretreated Φ40 mm × 5 mm M2 high-speed steel substrate was placed on a rotating holder in a vacuum chamber. The chamber was evacuated, and the baking current was gradually raised to 90 A for proper deposition temperature. When the substrate reached 120 °C and the chamber pressure dropped below 2.0×10^{-2} Pa, Ar was introduced at a flow rate of 12–13 sccm to initiate plasma cleaning, ensuring a clean substrate surface for subsequent coating deposition.

Metallic transition layer deposition followed to strengthen the coating–substrate bond for 5 min. Then, N_2 was introduced as a reactive gas. The N_2 flow rate was dynamically adjusted based on real-time pressure monitoring data to maintain a stable partial pressure of 0.38 Pa throughout the heating process as the furnace temperature was gradually increased from ambient to 800 °C at a controlled rate of 5 °C/min. The 15-min deposition process used a constant bias voltage of 180 V for a uniform and high-quality film.

4.4. Test Equipment

The surface and cross-sectional morphologies of the coatings were characterized using a ZEISS field emission scanning electron microscope (SEM) (Carl Zeiss Microscopy GmbH, Oberkochen, Germany) to analyze their microstructural features. The three-dimensional surface morphology and roughness of the coating were measured using a TaylorSure CCl2000 (Taylor Hobson Ltd., Leicester, UK) non-contact surface profilometer, providing a comprehensive assessment of the surface's microscopic geometry. The phase composition of the coating was analyzed using an XRD-7000 X-ray diffractometer (Shimadzu Corporation, Kyoto, Japan), with a monochromatic Al Kα (150 W, 15 kV, 1486.71 eV).

The critical load for coating adhesion was determined using a WS-2005 scratch adhesion tester (Zhuhai Lisite Science & Technology Co., Ltd., Zhuhai, China), with testing conditions including a preload of 0.1 N, a sliding speed of 5 mm/min, and a loading rate of 60 N/min. Each sample surface was tested three times, and the average value was recorded to ensure result reliability.

5. Discussion

5.1. Feature Set Analysis

The algorithm proposed in this study demonstrates significant distinctions from traditional modeling approaches. Conventional models typically rely solely on raw input data, such as composition and processing parameters, as input variables. However, in materials science, microstructural and mesoscale features, along with thermodynamic and kinetic properties, play critical roles in bridging the gap between raw inputs and target output properties [53–56]. To address this, this study employs an improved GA to optimize within the feature space, refining the molar ratios of elements in ternary metal nitrides. Based on the optimization results and selected features, SVR is utilized as the objective function to establish a quantitative relationship between elemental molar ratios and material hardness.

In this study, SVR is employed not only as the core objective function of the model but also to evaluate its generalization capabilities. To achieve this, cross-validation methods were applied using the dataset, and feature evaluations were conducted by calculating AUC (Area Under the Curve) values for all features listed in Table 1. AUC is a widely used metric for assessing the performance of binary classification models and is derived from the ROC (Receiver Operating Characteristic) curve. The ROC curve illustrates the trade-off between the True Positive Rate (TPR) and False Positive Rate (FPR) at various decision thresholds, providing an intuitive representation of how the model's classification performance changes with varying thresholds. In this context, higher-quality features result in ROC curves closer to the upper-left corner of the plot, indicating higher TPRs and lower FPRs.

Table 1. AUCS for detection performance with all features by ten times.

Times	1	2	3	4	5	6	7	8	9	10
AUC	0.945	0.901	0.899	0.915	0.963	0.959	0.891	0.932	0.964	0.941

The AUC value, representing the area under the ROC curve, ranges from 0 to 1 and quantifies the classification performance of the model. It can be interpreted as the probability that a randomly selected positive sample is ranked higher than a randomly selected negative sample by the model. An AUC value closer to 1 indicates superior classification performance, with the model accurately distinguishing between positive and negative samples. Conversely, an AUC value near 0.5 suggests performance akin to random guessing, reflecting poor classification ability. Through this evaluation method, the study not only validates the fitting capability of the SVR model but also comprehensively assesses its generalization performance in practical applications, providing robust quantitative support for further optimization.

$$TPR = \frac{TP}{TP + FN} \tag{27}$$

In this context, TP (True Positive) refers to the number of true positives, while FN (False Negative) refers to the number of false negatives.

$$FPR = \frac{FP}{FP + FN} \tag{28}$$

In this context, FP (False Positive) refers to the number of false positives, while TN (True Negative) refers to the number of true negatives.

In this study, the proposed model underwent a comprehensive performance evaluation using the aforementioned methodology, with its classification capability visualized through 10 sets of ROC curves (Figure 7). ROC curves closer to the upper-left corner of the plot indicate superior performance in distinguishing between positive and negative samples. By analyzing these curves, the model's classification effectiveness and predictive capability across varying decision thresholds were intuitively assessed.

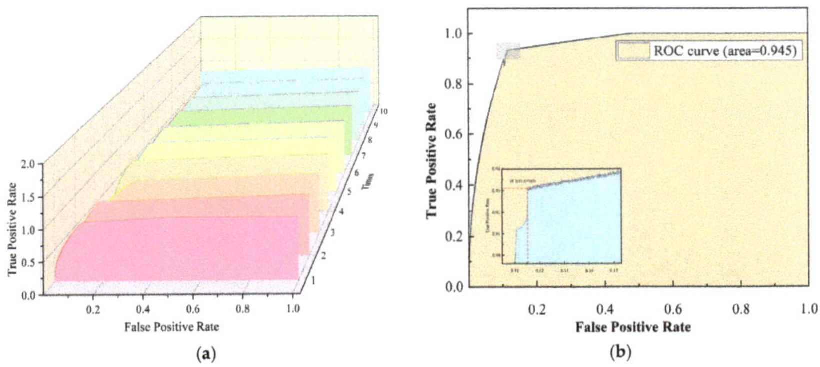

Figure 7. (**a**) Detection performance with all features by ten times. (**b**) The first-time detection performance with all features.

In the initial test, the model demonstrated high regression accuracy. Notably, by examining the inflection point of the ROC curve, the accuracy within the top 10% prediction range was estimated to be 90%. This result was further corroborated by the AUC values, underscoring the model's exceptional classification performance within this range. Specifically, the AUC quantifies the model's classification ability across different thresholds, with values closer to 1 indicating superior performance.

As summarized in Table 1, the AUC values from 10-fold cross-validation varied across experiments but yielded an average of 0.941. This average reflects a high degree of consistency in the model's classification performance across multiple tests, along with robust generalization capability. Overall, the near-unity AUC value further validates the model's high predictive accuracy and reliability in practical applications, highlighting its strong classification proficiency and remarkable stability.

This study employs a linear support vector machine (SVM) coefficient analysis method to quantitatively evaluate the importance of features within the dataset. To facilitate comparative analysis, each feature's importance coefficient was normalized to a relative value between 0 and 1, with the sum of all feature importance coefficients constrained to equal 1. A higher coefficient value indicates a more significant contribution of the feature to the model's predictive performance, reflecting its relative importance in the feature set.

As shown in Figure 8a, the analysis of feature importance coefficients reveals that subset 1 contains the highest number of features but has a lower average importance coefficient compared to subset 4. In contrast, although subset 4 comprises fewer features, its average importance coefficient is the highest. This finding indicates that the contributions of features within subset 1 are relatively distributed, while those in subset 4 make more pronounced contributions to the model's overall performance.

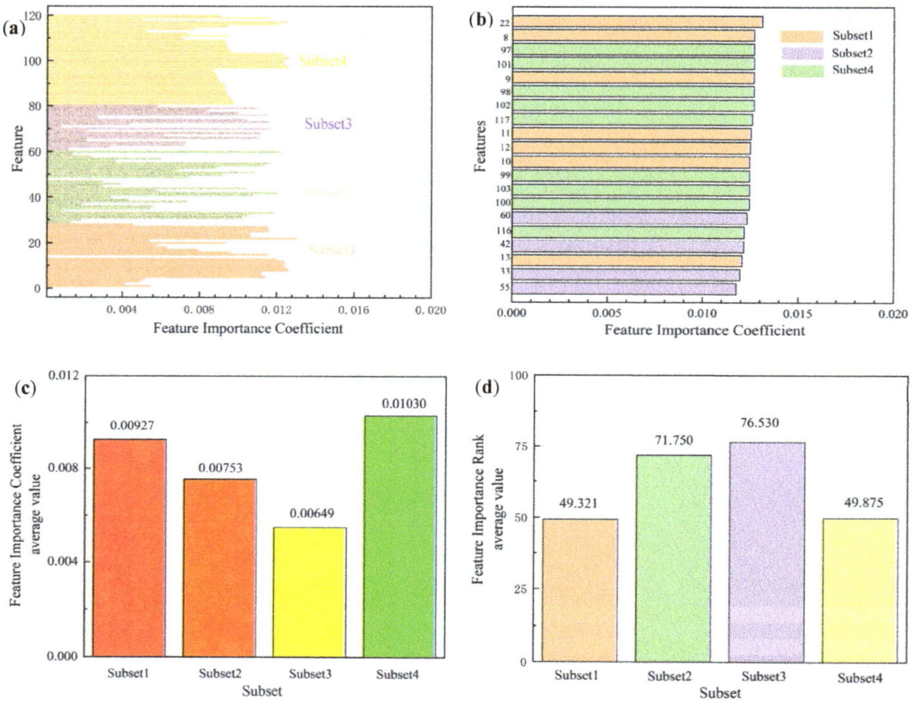

Figure 8. (**a**) The feature importance coefficient of all sets. (**b**) The feature importance coefficient of the top 20 in all sets. (**c**) The average value of the feature importance coefficient of all sets. (**d**) The average value of the feature importance coefficient rank of all sets.

In this study, we systematically assessed the relative importance of selected features in predicting material properties using SVR models. The experimental results demonstrate that the SVR model exhibits exceptional predictive accuracy, confirming its effectiveness and superiority for material property prediction tasks. To quantitatively analyze the contribution of each feature to the model's performance, we employed permutation testing to evaluate feature importance. The importance scores for each feature were normalized, ensuring values ranged from 0 to 1, with the sum of all feature importance scores equal to 1. The normalized importance scores reflect the relative contribution of each feature to the model's predictive performance, with higher scores indicating a more significant impact. The normalized feature importance of the model results is shown in Figure 8a.

A detailed analysis of the feature importance scores revealed that subset 4 had the highest average importance score of 0.0103 among all subsets, indicating its superior performance. Moreover, subset 4 occupied 9 out of the top 20 positions in the feature importance ranking (Figure 8b), outperforming the other subsets, with subset 1 following closely behind, occupying 7 positions. Further analysis indicated that the variance in feature importance rankings for subset 4 and subset 1 was significantly lower than for the other subsets (Figure 8d), suggesting that the contributions of these subsets were more stable. The features in subset 1 primarily relate to the atomic properties of metal elements and their crystal structures, while subset 4 focuses on the microscopic performance characteristics of metal nitrides. This result highlights a stronger correlation between the model's predictive performance, the fundamental properties of metal crystals, and the microscopic characteristics of metal nitrides, with the latter playing a pivotal role in hardness prediction.

In contrast, the features from subset 2 and subset 3 exhibited significantly lower importance scores, both in terms of average importance, the proportion of top 20 ranked features, and variance in importance scores, especially for subset 3, which showed markedly lower importance compared to the other subsets. This can be attributed to the fact that subset 3 features are primarily related to the macroscopic properties of elemental substances, which are less correlated with the performance of the final ternary nitrides. In comparison, subset 2 includes features related to periodic properties, ionization energy, and electronegativity, which contribute to the performance of ternary nitrides but have a lesser overall impact on predictive performance than those from subset 1 and subset 4.

In conclusion, feature selection should prioritize characteristics that reflect the microscopic properties of metal nitrides, rather than being limited to the atomic structure and crystal attributes of metal elements. This strategy not only enhances the predictive accuracy of the model but also provides crucial scientific insight into the intrinsic relationship between material properties and their microstructure, thereby advancing research in material design and performance optimization.

In this study, we employed an SVR model to investigate the relative importance of selected features in influencing the model's performance. The experimental results demonstrated that the SVR classifier achieved high detection accuracy, underscoring its effectiveness in predictive tasks. To further quantify the contribution of individual features to the model's performance, we utilized a permutation importance test to rank the features. The ranking results, illustrated in Figure 8b, indicate that Feature 1 and Feature 8 ranked the highest among all features, highlighting their dominant role in the prediction process and their substantial contribution to enhancing the model's detection accuracy.

In addition to evaluating the importance of individual features, we conducted a systematic analysis of feature interdependence, as correlations between features can significantly impact the model's predictive capabilities. To quantify the linear relationships between features, we employed the Pearson correlation coefficient (PCC). This metric evaluates the strength of the linear relationship between two variables, as expressed by Equation (29), where $\overline{f}_1$ and $\overline{f}_2$ represent the mean values of features f_1 and f_2, respectively. The Pearson correlation coefficient r ranges from −1 to 1, with values closer to −1 or 1 indicating stronger negative or positive linear correlations, respectively. A value of r = 0 signifies no linear correlation between the features.

$$r = \frac{\left\{\Sigma(f_1 - \overline{f}_1)(f_2 - \overline{f_2})\right\}}{\left\{\sqrt{\Sigma(f_1 - \overline{f}_1)^2}\sqrt{\Sigma(f_2 - \overline{f}_2)^2}\right\}} \tag{29}$$

Figure 9a,b present a visualization of the Pearson correlation coefficients between features, offering a clear depiction of their interrelationships. The analysis revealed significant correlations between certain features, suggesting the need to mitigate redundancy during the feature selection process. Redundant features can increase the complexity of the model and compromise its stability and generalizability. These findings emphasize the importance of addressing feature interdependencies to streamline the model structure and improve performance. The analysis of feature correlations thus provides critical theoretical insights for optimizing the model and enhancing its predictive capabilities.

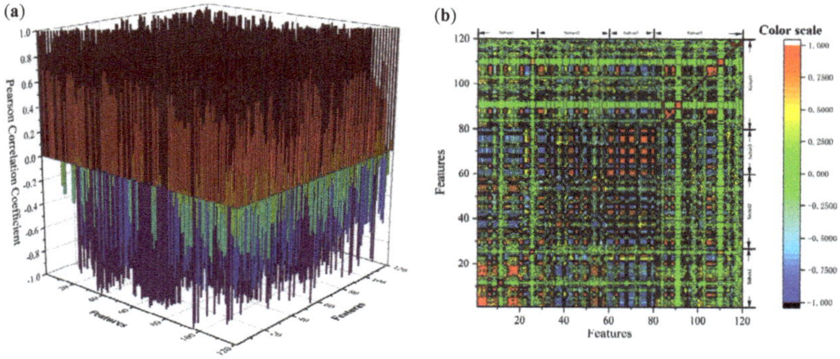

Figure 9. (**a**) The 3D plot of the pearson correlation coefficients between all features. (**b**) The 2D projection plot of the pearson correlation coefficients between all features.

From Figure 9a,b, it is evident that the features within subset 4 exhibit significantly lower correlations with those in other subsets. This can be attributed to the distinctive nature of subset 4 features, which are closely tied to the specific properties of metal nitrides. In contrast, the features of other subsets predominantly represent attributes related to elemental metals or individual elements. Metal nitrides exhibit unique physical and chemical characteristics at the microstructural and mesostructural levels, distinguishing them from the atomic structures and crystalline properties of elemental metals. Consequently, the structural and property-level dissimilarities between subset 4 and the other subsets result in reduced overlap and, thus, lower inter-subset correlations. This observation underscores the indispensable importance of subset 4 features in the model, particularly for tasks involving the prediction of properties specific to metal nitrides. The distinctiveness of subset 4 features is especially crucial for advancing our understanding of the unique characteristics of metal nitrides.

In contrast, features in subset 3 demonstrate higher correlations with those in other subsets, likely due to their focus on the macroscopic properties of materials. These macroscopic attributes often emerge as aggregated expressions of microstructural and mesostructural characteristics. For example, subset 3 features provide a holistic representation of material properties, integrating information from multiple scales. This integrative capacity accounts for the stronger correlations observed between subset 3 and the other subsets. Further analysis reveals that subset 3 plays a pivotal role in describing the overall performance of materials, while its high correlation with other subsets highlights the intrinsic connections between macroscopic properties and the underlying microstructural and mesostructural characteristics. This finding validates the model's capability to incorporate multi-scale features effectively, thereby establishing a robust foundation for comprehensive material property prediction.

5.2. Different Feature Subset Models

In traditional applications of materials machine learning, the inherent constitutive relationships between the physicochemical properties of materials, and their macroscopic mechanical performance are often excluded from modeling processes. As a result, materials design is frequently reduced to a purely mathematical or statistical optimization problem. However, such simplifications overlook the intrinsic mechanisms and complexities inherent to materials science. To address this limitation, the present study identifies and integrates a diverse set of feature variables that link materials design with their microstructural and mesoscale features as well as their thermodynamic and kinetic properties. By employing mathematical modeling, these physicochemical attributes are mapped onto a high-dimensional feature space within artificial intelligence frameworks, thereby playing a pivotal role in performance prediction.

The acquisition of raw input data in materials science is a time-intensive and costly endeavor, making data collection particularly challenging. This constraint results in limited sample spaces, which pose a significant bottleneck for the advancement of materials ML research. To overcome this challenge, this study introduces a high-dimensional feature space strategy, which not only effectively enriches the information content of available data but also improves data quality. This approach offers a novel solution to small dataset problems while substantially enhancing the model's capability to predict complex material properties.

To systematically investigate the influence of different feature subsets on the generalization ability of the SVR model, the study conducted a comprehensive analysis of all feature subsets. Specifically, by comparing the model's performance during training and testing with and without the inclusion of different feature subsets, the contribution of specific variables to the model's performance was assessed. The feature subsets used in the model are summarized in Table 2. The modeling process adhered strictly to the previously outlined methodologies to ensure the repeatability of the experiments and the reliability of the conclusions. This analysis provides critical insights for optimizing feature selection and advancing model performance, offering a robust foundation for future materials design efforts.

Table 2. The feature subsets used in the model.

Model	Feature Subset
GA-SVR-1	Subset 1 + subset 4
GA-SVR-2	Subset 2 + subset 3 + subset 4
GA-SVR-3	Subset 2 + subset 4
GA-SVR-4	Subset 1 + subset 3 + subset 4
GA-SVR-5	Subset 3 + subset 4
GA-SVR-6	Subset 1 + subset 2 + subset 4
GA-SVR-A	All sets (subset 1 + subset 2 + subset 3 + subset 4)

In addition to significantly enhancing the model's generalization capability, this approach offers an innovative solution to address the issue of small sample datasets. Traditionally, increasing the dataset size is a common strategy to mitigate small sample problems. However, in materials science, this process not only depends on the richness of the raw input data but also requires additional information obtained through standardized experiments, such as microstructural features. The time-consuming and costly nature of experimental processes, along with variations in experimental conditions that can compromise data consistency and quality, limits the applicability of this strategy in practice.

To overcome these challenges, this study introduces high-reliability variables related to the material itself, thereby significantly expanding the dimensionality of the dataset without relying on large-scale experimental data. Specifically, these variables provide more diverse and refined information for each sample, effectively broadening the feature space and enriching the hierarchical representation of the data. This strategy also mitigates the impact of experimental errors on data quality, improving data consistency and reliability. Moreover, by modeling within a high-dimensional feature space, the approach reduces the standard deviation during model training, alleviating the overfitting caused by insufficient data and significantly enhancing the robustness and reliability of the model's predictions.

This method transcends the limitations of traditional data augmentation techniques, offering a novel approach to addressing small sample issues and holding significant theoretical and practical value in improving model performance and optimizing material design.

The ultimate goal of the GA-SVR model developed in this study is to apply it to the design of novel materials. To explore the impact of incorporating different feature parameters on material design, we employed GA-SVR models based on various feature sets for the design of ternary metal nitride stainless steel. In these models, feature subsets were used as the objective function for the GA. By testing multiple GA-SVR models, we compared their performance across 10 test sets, evaluating metrics such as MAE, MSE, and R^2, with the results presented in Figure 10a. All SVR models achieved R^2 values greater than 87%, indicating that the models performed well across all feature combinations.

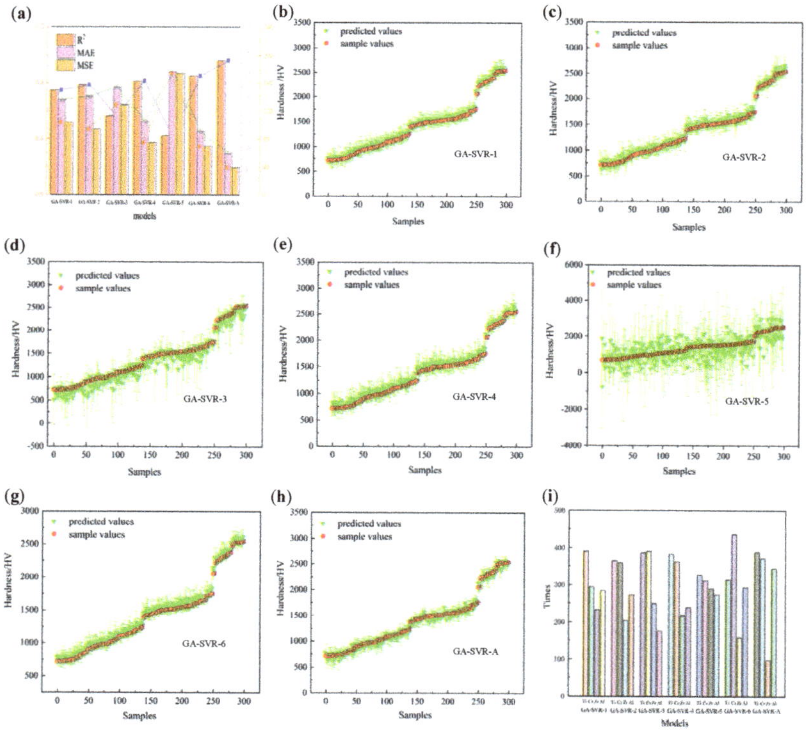

Figure 10. (**a**) The MSE, MAE, and R^2 of all models. (**b–h**) Ten prediction results of all models. (**i**) The times of element distribution of all models.

Further analysis of performance metrics, including MAE, MSE, and R^2 (Figure 10b), revealed that the GA-SVR-A model demonstrated significantly better fitting than the other models, with the most concentrated element distribution. This suggests that the contribution of feature subsets to the model is not simply a linear sum but rather a result of their mutual coupling and synergistic interaction. Additionally, models incorporating three feature subsets (such as GA-SVR-A) outperformed those with only two feature subsets, further confirming the importance of feature coupling in enhancing model performance. This also underscores the critical role of interactions between feature subsets in improving predictive accuracy.

Moreover, the comparison of model performances indicates that GA-SVR-6 and GA-SVR-4 follow GA-SVR-A in terms of fitting accuracy. The shared characteristic of these models is that they include all three feature subsets and specifically feature subset 1. This finding aligns closely with the analysis of feature subset importance and correlation in Section 5.1, providing further evidence that the coupling between feature subsets is key to improving model prediction accuracy. Consequently, this study suggests that in material design tasks, feature selection should focus not only on the individual effects of features but also on the interactions between feature subsets.

In traditional experimental designs, the formation process of substitutional solid solutions is also considered, involving the substitution of different metal atoms in the lattice. This process is influenced by factors such as diffusion, temperature, lattice defects, and atomic size differences. In Ti-, Cr-, Zr-, and Al-based multi-metal nitrides, the kinetics of substitutional solid solution formation determine the microstructure of the nitrides. For example, the substitution of Ti and Zr atoms in the lattice can enhance the lattice stability, thus affecting the material's hardness and toughness. The diffusion rate is a key parameter controlling the formation of substitutional solid solutions. Additionally, lattice defects such as vacancies and dislocations serve as fast channels for diffusion, accelerating the migration of solute atoms and promoting the formation of substitutional solid solutions. However, these defects may also introduce stress concentration, leading to increased brittleness in the material. These issues are typically considered by designers based on experience.

As shown in Figure 10a, the GA-SVR-3 model exhibits slightly lower MAE, MSE, and R^2 values compared to GA-SVR-4, but these values are still higher than those of GA-SVR-4. Notably, the MAE, MSE, and R^2 values of GA-SVR-6 are significantly better than those of the other five models, except for GA-SVR-A. This improvement can be attributed to the inclusion of feature variables in subset 2, which consist of atomic structure properties (including atomic structure, atomic energy, and chemical bonds) as well as thermodynamic and kinetic properties such as Gibbs free energy and entropy. These properties govern the structural characteristics, phonon density of states, and Debye temperature of multi-metal nitrides, thereby influencing their macroscopic mechanical properties. Furthermore, these thermodynamic properties affect the stability of new structures and the rate of diffusion. However, due to the steady-state deposition of coatings in multi-arc ion plating experiments, the influence of kinetics is greatly reduced. Nevertheless, the unique atomic properties collectively shape the microstructure and chemical stability of nitrides, thereby determining the material's mechanical properties. For example, Ti, Cr, Zr, and Al, all transition metals, have distinct electronic configurations and atomic radii. Ti and Zr share similar electronic structures (d^2sp^3 hybridization), enabling them to form strong metal–nitrogen bonds in nitrides. Cr, with d^3sp^3 hybridization, possesses higher electronic density, which facilitates the formation of stable nitride structures. Al, though not a transition metal, forms stable chemical bonds with nitrogen due to its sp^2 hybridization. In terms of atomic energy, the differences in atomic energy between Ti, Cr, Zr, and Al result in different behaviors in nitrides. Ti and Zr have higher melting and boiling points, indicating better stability at high temperatures. Although chromium has a lower melting point and boiling point, its higher hardness and corrosion resistance make it an ideal component for enhancing the hardness of nitrides. Aluminum's low melting point and high thermal conductivity improve the processing performance of nitrides. Regarding chemical bonds, the strength and nature of metal–nitrogen bonds play a crucial role in the hardness of nitrides. The strong metal–nitrogen bonds in Ti and Zr contribute to higher hardness and wear resistance, while Cr forms moderate-strength bonds that balance hardness and toughness. Aluminum forms relatively weaker bonds, but its inclusion improves the oxidation resistance and thermal stability of the nitrides.

In contrast, the GA-SVR-5 model shows the lowest performance. The feature subset 3 focuses on the macroscopic properties of each elemental material. In traditional experiments, macroscopic attributes such as hardness are often linearly correlated to estimate the total contribution of the final result. However, in actual material preparation, many factors contribute to the final result, and the relationship with macroscopic properties is often nonlinear. Compared to this, SVR models can significantly reduce preparation complexity. However, macroscopic features usually provide an overall description of material properties and may not capture the microscopic mechanisms that influence material hardness. This is likely the reason for the poorer performance of the GA-SVR-5 model.

From the perspective of element distribution (as shown in Figure 10i), it is observed that models incorporating subset 2 feature subsets show a more concentrated element distribution, with Ti and Cr elements appearing much more frequently than others. This indicates a complex relationship between the hardness of metal nitrides and their lattice distortions, particularly with respect to chemical bonds and crystal structures. Further analysis reveals that elements' electronegativity, ionization energy, and thermodynamic properties make significant positive contributions to the hardness prediction of metal nitrides. These physicochemical properties directly affect the strength of metal nitrides, mainly through their role in modulating atomic bonding forces and influencing crystal structure stability. Specifically, elements such as Ti and Cr, with high electronegativity and good thermodynamic stability, contribute to nitrides with higher hardness and stronger resistance to deformation. Thus, the electronegativity, ionization energy, and related thermodynamic properties of these elements play a crucial role in predicting the hardness of metal nitrides, revealing the intrinsic connection between the microstructure and macroscopic mechanical properties of these materials.

5.3. Analysis of Prediction Results and Experimental Validation

In this study, the GA-SVR-A model was used to predict hardness, and the predictions were compared with experimental data. The test results demonstrated that the GA-SVR-A model significantly outperformed other models in terms of MAE, MSE, and R^2. The model's superiority lies not only in the linear combination of the four feature subsets but also in its ability to provide complementary information between subsets. This reduces the model's over-reliance on any particular feature set, thereby enhancing its generalization capability.

For hardness prediction, the GA-SVR-A model was tested with 100 predictions, resulting in an average predicted hardness of 2960.02 HV. Notably, 82 of the predictions exceeded the maximum hardness value in the dataset. To further explore the relationship between element molar ratios and hardness in the predictive model, we selected the element ratio from the first prediction as a baseline and performed variance analysis on the remaining predictions against this baseline. This approach quantifies the differences in element composition across predictions, allowing for an assessment of the impact of element molar ratios on hardness prediction. The specific formula used is shown in Equation (30).

$$VM = \frac{1}{n}\sum (x_i - x_1)^2 \quad (i = 2, \ldots\ldots, 100) \tag{30}$$

Here, VM represents the variance between element molar ratios, x_i denotes the molar ratio of elements in the i-th prediction, and x_1 represents the molar ratio in the first prediction.

To mitigate the impact of preparation accuracy and measurement errors on the model's predictions, clustering analysis was performed on samples with hardness predictions exceeding 3200 HV. The specific steps are as follows:

(1) Calculation of distances between samples: First, the distances between the 19 samples with hardness values greater than 3200 HV were calculated. The average of these distances was then obtained using Equation (31) to provide an overall distribution of the sample points.

$$R_{nei} = \frac{1}{4n}\sqrt{(x_i - x_j)^2 + (y_i - y_j)^2}(i, j = 1, 2, 3, \ldots 19 \; and \; i \neq j) \quad (31)$$

(2) Clustering analysis: The DBSCAN algorithm was employed to cluster the 19 sample points. The neighborhood radius was set to R_{nei}, with the minimum sample number $MinPts$ set to 3. If a point had a neighborhood sample count greater than R_{nei} but fewer than MinPts, it was considered a noise point and excluded from the clustering process. This method effectively identifies and removes outliers, thereby enhancing the accuracy of the clustering analysis.

(3) Objective function construction and gradient descent optimization: An objective function was established to compute the centroid coordinates (a,b) and the minimum radius r_{area} for each cluster, with the clustering areas labeled as Area 1, Area 2, and Area 3. This approach effectively extracts core data features within the clustering regions.

$$J(a, b, r_{area}) = \sum_{i=1}^{19} \left((x_i - a)^2 + (y_i - b)^2 - {r_{area}}^2\right)^2 \quad (32)$$

Within the clustering regions, it was assumed that the relationship between material hardness and molar ratio followed the functional relationship shown in Figure 11b. Therefore, the element ratios extracted within each neighborhood should correspond to the hardness values within that region. Based on this assumption, the value ranges for the three clustering areas are shown in Table 1. To validate the model's predictive accuracy, five experimental sets were conducted within each of the three areas, and the corresponding molar ratios were recorded (see Table 3). The experimental results indicated that Area 2 contained the most samples, followed by Area 1, with Area 3 containing the fewest samples.

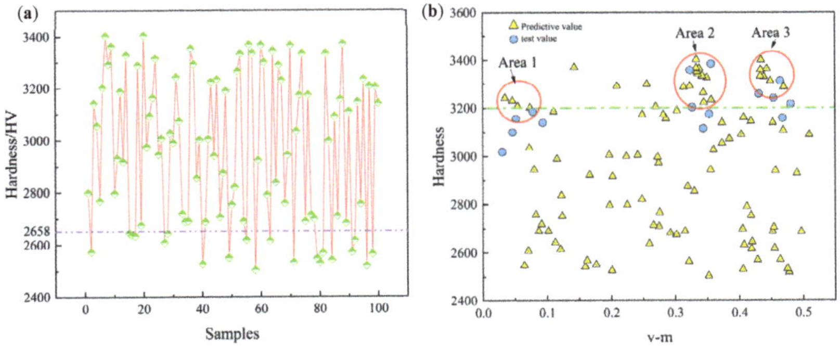

Figure 11. (**a**) Ten prediction results of the GA-SVR-A model. (**b**) Predicted and experimental values of the GA-SVR-A model.

Table 3. Range of metal content proportions observed in the three areas of Figure 11.

Area	Ti	Al	Cr	Zr
Area 1	0	0.0855~0.125	0.106~0.153	0.218~0.290
Area 2	0.0865~0.115	0.189~0.269	0.187~0.215	0
Area 3	0.218~0.255	0.185~0.166	0.147~0.181	0

To validate the model's predictions, five sets of experiments were conducted within each of the three regions, with molar ratios provided in Table 4. In all validation experiments, all sample points in Area 2 fell within the region. In contrast, only three sample points from Area 3 and one from Area 1 were found within their respective regions. Subsequently, all fifteen samples were input into the trained GA-SVR-A model, and the results showed that the samples from Area 2 exhibited the smallest prediction errors. This suggests that the more prediction points within a region, the closer the relationship between the element molar ratio and hardness aligns with the functional relationship shown in Figure 11b. Conversely, the fewer the prediction points within a region, the greater the deviation from the relationship.

Table 4. Predicted and experimental values of the GA-SVR-A model.

Area	Experimental Sample	Element Ratio	Test Strength (HV)	Model Prediction Strength (HV)	Prediction Accuracy
Area 1	Area1-XN1	$Al_{0.08}Cr_{0.13}Zr_{0.22}N_{0.5}$	3189	3230	98.73%
	Area1-XN2	$Al_{0.06}Cr_{0.06}Zr_{0.13}N_{0.5}$	3160	3164	99.87%
	Area1-XN3	$Al_{0.085}Cr_{0.15}Zr_{0.265}N_{0.5}$	3140	3244	96.79%
	Area1-XN4	$Al_{0.10}Cr_{0.12}Zr_{0.22}N_{0.57}$	3100	3253	95.30%
	Area1-XN5	$Al_{0.10}Cr_{0.145}Zr_{0.26}N_{0.5}$	3017	3241	93.09%
Area 2	Area2-XN1	$Ti_{0.1q}Al_{0.19}Cr_{0.21}N_{0.5}$	3356	3293	98.12%
	Area2-XN2	$Ti_{0.09}Al_{0.20}Cr_{0.21}N_{0.5}$	3207	3348	95.79%
	Area2-XN3	$Ti_{0.09}Al_{0.17}Cr_{0.18}N_{0.57}$	3178	3362	94.53%
	Area2-XN4	$Ti_{0.095}Al_{0.20}Cr_{0.2}N_{0.5}$	3385	3271	96.63%
	Area2-XN5	$Ti_{0.11}Al_{0.19}Cr_{0.2}N_{0.5}$	3115	3398	91.67%
Area 3	Area3-XN1	$Ti_{0.21}Al_{0.26}Cr_{0.28}N_{0.25}$	3314	3241	97.79%
	Area3-XN2	$Ti_{0.13}Al_{0.04}Cr_{0.08}N_{0.5}$	3260	3292	99.03%
	Area3-XN3	$Ti_{0.15}Al_{0.27}Cr_{0.33}N_{0.25}$	3243	3287	98.66%
	Area3-XN4	$Ti_{0.23}Al_{0.10}Cr_{0.17}N_{0.5}$	3220	3291	97.84%
	Area3-XN5	$Ti_{0.23}Al_{0.08}Cr_{0.17}N_{0.5}$	3162	3282	96.34%

The XRD spectrum of the sample Area2-XN5 also shows four main diffraction peaks, as illustrated in Figure 12a, corresponding to the (111), (200), and (220) planes, and the CrTiAl alloy phase diffraction peak, all exhibiting typical face-centered cubic lattice characteristics. From the surface morphology in Figure 12b, it is evident that the coating surface is continuous; dense; and free from peeling, porosity, or cracking. The alloy phase is observed to originate from droplets within the coating. The distribution of elements in the coating is shown in Figure 12c,d,f,g. These results are in line with our original design concept.

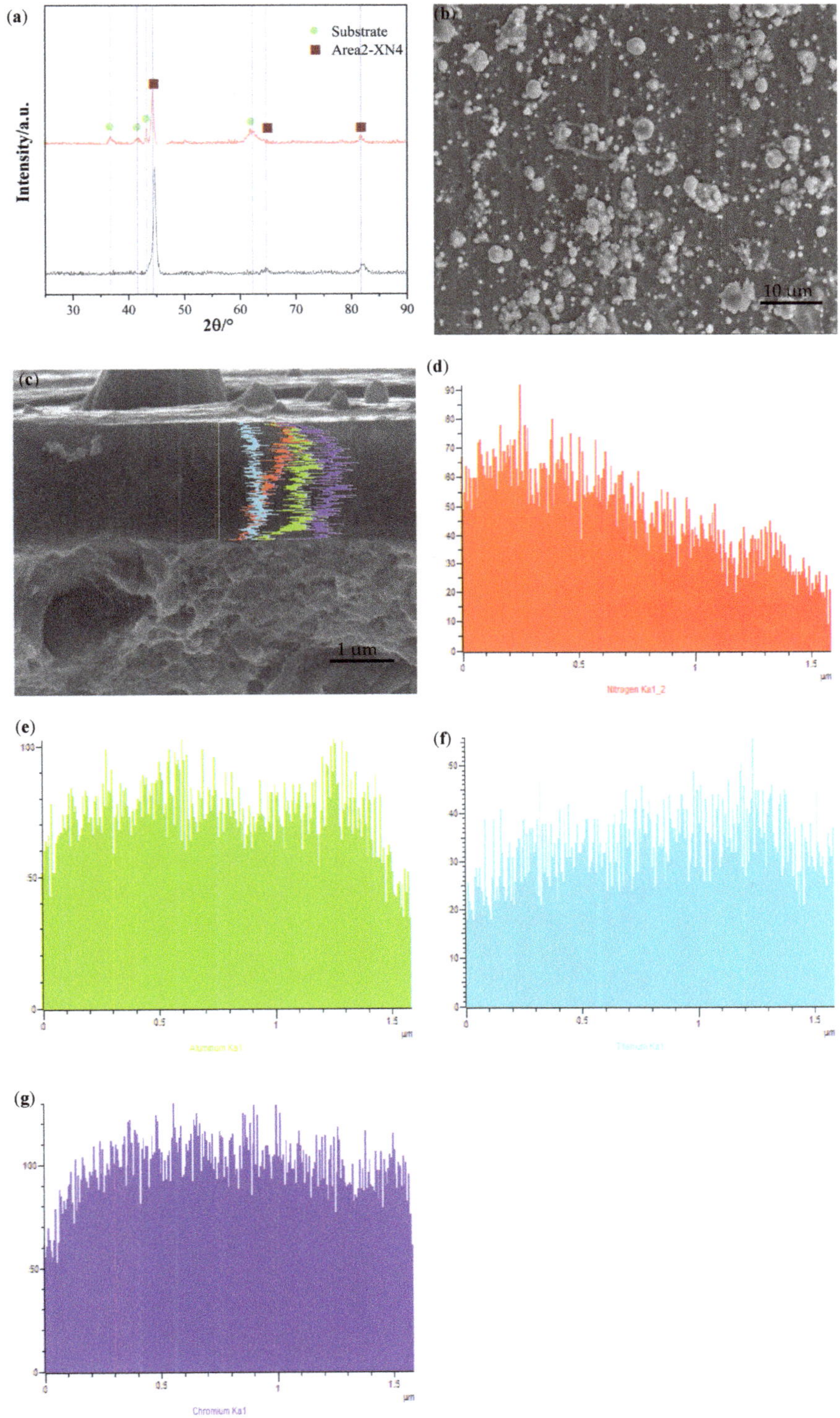

Figure 12. (**a**) The XRD of Area2-XN4. (**b**) TEM image of Area2-XN4. (**c**–**g**) EBSD of Area2-XN4.

6. Conclusions

In this work, a novel approach is presented for the design of multicomponent metal nitride hard coatings. The method begins with the electronic structure of high-hardness unary metal nitrides, establishing a compositional search space based on the experimental conditions. This search space is further constrained by the crystal structure and strengthening mechanisms, and a target function is defined for the model. An improved GA combined with SVR is then employed to design a high-hardness multicomponent metal nitride, which is experimentally validated through hardness testing.

The introduction of feature sets enhances the model's ability to consider strengthening mechanisms across micro-, macro-, and mesoscopic dimensions. To investigate the influence of different dimensional strengthening mechanisms on hardness enhancement, the study reveals that subset 1 has a significantly higher impact than other subsets. This is attributed to the inclusion of higher-dimensional features, which improve the model's ability to capture complex relationships between variables. Additionally, the inclusion of such high-dimensional features is particularly beneficial for systems with small datasets, where traditional methods may be less effective.

Experimental validation confirms that the predicted optimal parameters align closely with the final performance of the coatings, demonstrating the robustness and accuracy of the proposed model. The use of clustering analysis minimizes experimental and data errors, further enhancing the reliability of the predictions. Of the 100 predictions made by the model, 82 were higher than the maximum hardness in the dataset, and the prediction accuracy for the best sample was 91.6%.

Despite these achievements, the approach has limitations that warrant future investigation. The dataset of 233 entries is relatively small, potentially limiting the model's ability to capture the full compositional variability of multicomponent nitrides. Additionally, the model's reliance on multi-arc ion plating for validation may restrict its generalizability to other synthesis methods, such as magnetron sputtering. Furthermore, feature engineering based on solid solution strengthening theory assumes specific physical mechanisms, which may not fully account for all interactions in complex nitride systems. Future work should focus on expanding the dataset with diverse nitride compositions, testing the model across various deposition techniques, and refining feature engineering by incorporating additional physical models to enhance predictive robustness.

Supplementary Materials: The following supporting information can be downloaded at: https://www.mdpi.com/article/10.3390/ma18153478/s1, Table S1. The elements of the set subset-N. Table S2. All the features and the details of their subsets. Table S3. List of Abbreviations.

Author Contributions: Y.G.: Conceptualization, Methodology, Writing—Original Draft, Supervision; J.W.: Methodology, Software, Writing—Review & Editing; Y.Z.: Formal Analysis, Writing—Original Draft, Visualization; B.D.: Investigation, Data Curation, Validation; G.L.: Investigation, Resources, Project Administration; J.Z.: Validation, Data Curation, Visualization; L.B.: Resources, Funding Acquisition; R.L.: Resources, Project Administration, Funding Acquisition. Y.L.: Formal Analysis, Data Curation; Validation, Writing—Review & Editing. All authors have read and agreed to the published version of the manuscript.

Funding: This work was financially supported by the National Natural Science Foundation of China (Grant No. 52005432, Grant No. 52405382) and Hebei Natural Science Foundation of China (No. E2021203179).

Institutional Review Board Statement: Not applicable.

Informed Consent Statement: Not applicable.

Data Availability Statement: The original contributions presented in this study are included in the article/Supplementary Materials. Further inquiries can be directed to the corresponding author.

Conflicts of Interest: The authors declare no conflict of interest.

References

1. Mohan, K.S.S.; Gunasekaran, S.; Manjubashini, D.; Umayal, S.; Sivaranjani, S.; Subramanian, B. Metal Nitride Coatings of (Cr/Mo)N and (Cr/Ti)N by Ion-Assisted Co-sputtering for Corrosion-Resistant Applications. *J. Mater. Eng. Perform.* **2024**, *33*, 10614–10622. [CrossRef]
2. Li, J.C.; Chen, Y.J.; Zhao, Y.M.; Shi, X.; Wang, S.; Zhang, S. Super-hard (MoSiTiVZr)Nx high-entropy nitride coatings. *J. Alloys Compd.* **2022**, *926*, 166807. [CrossRef]
3. Park, H.J.; Kim, Y.S.; Lee, Y.H.; Hong, S.H.; Kim, K.S.; Park, Y.K.; Kim, K.B. Design of nano-scale multilayered nitride hard coatings deposited by arc ion plating process: Microstructural and mechanical characterization. *J. Mater. Res. Technol.* **2021**, *15*, 572–581. [CrossRef]
4. Ghailane, A.; Makha, M.; Larhlimi, H.; Alami, J. Design of hard coatings deposited by HiPIMS and dcMS. *Mater. Lett.* **2020**, *280*, 128540. [CrossRef]
5. Kumar, D.D.; Kumar, N.; Kalaiselvam, S.; Dash, D.S.; Jayavel, A.R. Wear resistant super-hard multilayer transition metal-nitride coatings. *Surf. Interfaces* **2017**, *7*, 74–82. [CrossRef]
6. Chang, Y.Y.; Weng, S.Y.; Chen, C.H.; Fu, F.X. High temperature oxidation and cutting performance of AlCrN, TiVN and multilayered AlCrN/TiVN hard coatings. *Surf. Coat. Technol.* **2017**, *332*, 494–503. [CrossRef]
7. Wang, S.M.; Antonio, D.; Yu, X.H.; Zhang, J.; Cornelius, A.L.; He, D.; Zhao, Y. The Hardest Superconducting Metal Nitride. *Sci. Rep.* **2015**, *5*, 13733. [CrossRef] [PubMed]
8. Li, X.L.; Jin, G.; Liu, F.; Sun, Y.F.; Guan, S.K.; Chen, Y. Simultaneous enhancements in strength and ductility of hexagonal BN nanosheet reinforced AlSi10Mg composites. *J. Alloys Compd.* **2024**, *992*, 174577. [CrossRef]
9. Chen, Y.C.; Fan, F.Y.; Tsai, M.H.; Wu, C.H.; Chen, S.T.; Ou, S.F. Microstructure and mechanical properties of Ti nitride/Ni metal-based composites fabricated by reactive sintering. *Ceram. Int.* **2019**, *45*, 10834–10839. [CrossRef]
10. Tao, Y.; Ke, G.S.; Xie, Y.; Chen, Y.G.; Shi, S.Q.; Guo, H.B. Adhesion strength and nucleation thermodynamics of four metals (Al, Cu, Ti, Zr) on AlN substrates. *Appl. Surf. Sci.* **2015**, *357*, 8–13. [CrossRef]
11. Zhang, Z.L.; Ren, Y.X.; He, G.Y.; Yang, Z.F. Enhancing erosion resistance of ceramic/metal multilayer coating by crack inhibition design. *J. Vac. Sci. Technol. A* **2024**, *42*, 053107. [CrossRef]
12. Zhang, M.C.; Ai, X.; Liang, X.; Chen, H.; Zou, X.X. Key Role of Local Chemistry in Lattice Nitrogen-Participated N2-to-NH_3Electrocatalytic Cycle over Nitrides. *Adv. Funct. Mater.* **2023**, *33*, 2306358. [CrossRef]
13. Tang, L.L.; Yao, D.X.; Xia, Y.F.; Liang, H.Q.; Zeng, Y.P. Effect of interfacial microstructure evolution on the peeling strength and fracture of silicon nitride/oxygen-free copper foil joints brazed with Ag-Cu-TiH_2 filler. *J. Eur. Ceram. Soc.* **2023**, *43*, 4374–4385. [CrossRef]
14. Chao, Q.; Mateti, S.; Annasamy, M.; Imran, M.; Joseph, J.; Cai, Q.R.; Li, L.H.; Cizek, P.; Hodgson, P.D.; Chen, Y.; et al. Nanoparticle-mediated ultra grain refinement and reinforcement in additively manufactured titanium alloys. *Addit. Manuf.* **2021**, *46*, 102173. [CrossRef]
15. Grosso, B.F.; Davies, D.W.; Zhu, B.A.; Walsh, A.; Scanlon, D.O. Accessible chemical space for metal nitride perovskites. *Chem. Sci.* **2023**, *14*, 9175–9185. [CrossRef] [PubMed]
16. Sun, W.H.; Bartel, C.J.; Arca, E.; Bauers, S.R.; Matthews, B.; Orvañanos, B.; Chen, B.R.; Toney, M.F.; Schelhas, L.T.; Tumas, W.; et al. A map of the inorganic ternary metal nitrides. *Nat. Mater.* **2019**, *18*, 732–752. [CrossRef] [PubMed]
17. Sun, W.H.; Holder, A.; Orvañanos, B.; Arca, E.; Zakutayev, A.; Lany, S.; Ceder, G. Thermodynamic Routes to Novel Metastable Nitrogen-Rich Nitrides. *Chem. Mater.* **2017**, *29*, 6936–6946. [CrossRef]
18. Owolabi, T.O.; Akande, K.O.; Olatunji, S.O. Estimation of average surface energies of transition metal nitrides using computational intelligence technique. *Soft Comput.* **2017**, *21*, 6175–6182. [CrossRef]
19. Junrear, J.; Sakunasinha, P.; Chiangga, S. The Optimization of Metal Nitride Coupled Plasmon Waveguide Resonance Sensors Using a Genetic Algorithm for Sensing the Thickness and Refractive Index of Diamond-like Carbon Thin Films. *Photonics* **2022**, *9*, 332. [CrossRef]
20. Zhang, C.; Sun, Y.; Zhang, F.; Ho, K.M.; Wang, C.Z. An ultra-incompressible Mn_3N compound predicted by first-principles genetic algorithm. *J. Appl. Phys.* **2020**, *128*, 055112. [CrossRef]
21. Fang, S.F.; Wang, M.P.; Qi, W.H.; Zheng, F. Hybrid genetic algorithms and support vector regression in forecasting atmospheric corrosion of metallic materials. *Comput. Mater. Sci.* **2008**, *44*, 647–655. [CrossRef]

22. Abdesselam, Y.; Rezgui, I.; Naoun, M.; Belloufi, A.; Mezoudj, M.; Zerrouki, D. Experimental investigation and optimization of manufacturing processes of Ni-P-Y_2O_3composite coatings by multiple linear regression method based on genetic algorithm. *Int. J. Adv. Manuf. Technol.* **2023**, *9*, 3995–4019. [CrossRef]
23. Metni, H.; Takeuchi, I.; Stanev, V. Predicting the superconducting critical temperature in transition metal carbides and nitrides using machine learning. *Phys. C Supercond. Appl.* **2023**, *605*, 1354209. [CrossRef]
24. Qian, C.F.; Sun, K.W.; Bao, W.Z. Recent advance on machine learning of MXenes for energy storage and kandaoconversion. *Int. J. Energy Res.* **2022**, *46*, 21511–21522. [CrossRef]
25. Denny, S.R.; Lin, Z.X.; Porter, W.N.; Artrith, N.; Chen, J.G. Machine learning prediction and experimental verification of Pt-modified nitride catalysts for ethanol reforming with reduced precious metal loading. *Appl. Catal. B Environ.* **2022**, *312*, 121380. [CrossRef]
26. Ohkubo, I.; Hou, Z.; Lee, J.N.; Aizawa, T.; Lippmaa, M.; Chikyow, T.; Tsuda, K.; Mori, T. Realization of closed-loop optimization of epitaxial titanium nitride thin-film growth via machine learning. *Mater. Today Phys.* **2021**, *16*, 100296. [CrossRef]
27. Xia, K.; Gao, H.; Liu, C.; Yuan, J.N.; Sun, J.; Wang, H.T.; Xing, D.Y. A novel superhard tungsten nitride predicted by machine-learning accelerated crystal structure search. *Sci. Bull.* **2018**, *63*, 817–824. [CrossRef] [PubMed]
28. Tian, Y.J. The hardest transition metal nitride predicted from machine learning. *Sci. Bull.* **2018**, *63*, 947–948. [CrossRef] [PubMed]
29. Yang, Z.; Ren, Y.; Zhang, Y.; Zhang, Z.L.; He, G.Y.; Zhang, Z.L. Structure Optimization and Failure Mechanism of Metal Nitride Coatings for Enhancing the Sand Erosion Resistance of Aluminum Alloys. *Coatings* **2023**, *13*, 2074. [CrossRef]
30. Li, X.; Xin, W.; Zheng, X.; Ren, Z.A.; Sun, D.Q.; Lu, W.L. Microstructural Characterization and Formation Mechanism of Nitrided Layers on Aluminum Substrates by Thermal Plasma Nitriding. *Metals* **2019**, *9*, 523. [CrossRef]
31. Batsanov, S.S. Energy Electronegativity and Chemical Bonding. *Molecules* **2022**, *27*, 8251. [CrossRef] [PubMed]
32. Mizoguchi, T.; Sasaki, T.; Tanaka, S.; Matsunaga, K.; Yamamoto, T.; Kohyama, M.; Ikuhara, Y. Chemical bonding, interface strength, and oxygen K electron-energy-loss near-edge structure of the Cu/Al_2O_3 interface. *Phys. Rev. B* **2006**, *74*, 235408. [CrossRef]
33. Johnson, D.D. Electronic Basis of the Strength of Materials. *Phys. Today* **2004**, *57*, 56–57. [CrossRef]
34. Freiesleben, T.; König, G.; Molnar, C.; Tejero-Cantero, A. Scientific Inference with Interpretable Machine Learning: Analyzing Models to Learn About Real-World Phenomena. *Minds Mach.* **2024**, *34*, 32. [CrossRef]
35. Walbrühl, M.; Linder, D.; Ågren, J.; Borgenstam, A. Modelling of solid solution strengthening in multicomponent alloys. *Mater. Sci. Eng. A* **2017**, *700*, 301–311. [CrossRef]
36. Lugovy, M.; Slyunyayev, V.; Brodnikovskyy, M. Solid solution strengthening in multicomponent fcc and bcc alloys: Analytical approach. *Prog. Nat. Sci. Mater. Int.* **2021**, *31*, 95–104. [CrossRef]
37. Jacob, K.T.; Raj, S.; Rannesh, L. Vegard's law: A fundamental relation or an approximation? *Int. J. Mater. Res.* **2007**, *98*, 776–779. [CrossRef]
38. Bullard, J.W.; Scherer, G.W. An Ideal Solid Solution Model for C-S-H. *J. Am. Ceram. Soc.* **2016**, *99*, 4137–4145. [CrossRef] [PubMed]
39. Myhill, R.; Connolly, J.A.D. Notes on the creation and manipulation of solid solution models. *Contrib. Mineral. Petrol.* **2021**, *176*, 86. [CrossRef]
40. Goldschmidt, H.J. Interstitial Solid Solutions. In *Interstitial Alloys*; Goldschmidt, H.J., Ed.; Springer: Boston, MA, USA, 1967; pp. 60–87.
41. Liu, C.; Lu, W.; Xia, W.; Du, C.W.; Rao, Z.Y.; Best, J.P.; Brinckmann, S.; Lu, J.; Gault, B.; Dehm, G.; et al. Massive interstitial solid solution alloys achieve near-theoretical strength. *Nat. Commun.* **2022**, *13*, 1102. [CrossRef] [PubMed]
42. Kadulkar, S.; Sherman, Z.M.; Ganesan, V.; Truskett, T.M. Machine Learning-Assisted Design of Material Properties. *Annu. Rev. Chem. Biomol. Eng.* **2022**, *13*, 235–254. [CrossRef] [PubMed]
43. Liu, Y.L.; Niu, C.; Wang, Z.; Gan, Y.; Zhu, Y.; Sun, S.H.; Shen, T. Machine learning in materials genome initiative: A review. *J. Mater. Sci. Technol.* **2020**, *57*, 113–122. [CrossRef]
44. Di Schino, A.; Testani, C. Microstructure and Properties in Metals and Alloys. *Metals* **2023**, *13*, 1320. [CrossRef]
45. Dolzhenko, A.; Tikhonova, M.; Kaibyshev, R.; Belyakov, A. Microstructures and Mechanical Properties of Steels and Alloys Subjected to Large-Strain Cold-to-Warm Deformation. *Metals* **2022**, *12*, 454. [CrossRef]
46. Wang, J.; Xiao, X. Thermodynamic, Kinetic and Strength Calculation of High Zinc Containing Al-Zn-Mg-Cu Alloys. *Crystals* **2022**, *12*, 181. [CrossRef]
47. Heaton, J. An empirical analysis of feature engineering for predictive modeling. In Proceedings of the SoutheastCon 2016, Norfolk, VA, USA, 30 March–3 April 2016; pp. 1–6.
48. Shi, M.L.; Lv, L.Y.; Guo, Z.G.; Sun, W.; Song, X.G.; Li, H.Y. High-Low Level Support Vector Regression Prediction Approach (HL-SVR) for Data Modeling with Input Parameters of Unequal Sample Sizes. *Int. J. Comput. Methods* **2021**, *18*, 2150029. [CrossRef]
49. Nguyen, H.; Choi, Y.; Bui, X.N.; Trung, N.T. Predicting Blast-Induced Ground Vibration in Open-Pit Mines Using Vibration Sensors and Support Vector Regression-Based Optimization Algorithms. *Sensors* **2020**, *20*, 132. [CrossRef] [PubMed]

50. Peng, C.; Che, Z.Y.; Liao, T.W.; Zhang, Z.W. Prediction using multi-objective slime mould algorithm optimized support vector regression model. *Appl. Soft Comput.* **2023**, *145*, 110580. [CrossRef]
51. Akour, M.; Alenezi, M.; Alsghaier, H. Software Refactoring Prediction Using SVM and Optimization Algorithms. *Processes* **2022**, *10*, 1611. [CrossRef]
52. Tao, P.Y.; Sun, Z.; Sun, Z.X. An Improved Intrusion Detection Algorithm Based on GA and SVM. *IEEE Access* **2018**, *6*, 13624–13631. [CrossRef]
53. Cox, T.; Motevalli, B.; Opletal, G.; Barnard, A.S. Feature Engineering of Solid-State Crystalline Lattices for Machine Learning. *Adv. Theory Simul.* **2020**, *3*, 1900190. [CrossRef]
54. Cisty, M.; Danko, M.; Kohnova, S.; Povazanová, B.; Trizna, A. Machine Learning Enhanced by Feature Engineering for Estimating Snow Water Equivalent. *Water* **2024**, *16*, 2285. [CrossRef]
55. Wei, J.; Chu, X.; Sun, X.Y.; Xu, K.; Deng, H.X.; Chen, J.G.; Wei, Z.M.; Lei, M. Machine learning in materials science. *InfoMat* **2019**, *1*, 338–358. [CrossRef]
56. Wagner, N.; Rondinelli, J.M. Theory-guided Machine learning in Materials science. *Front. Mater.* **2016**, *3*, 28. [CrossRef]

materials

MDPI

Article

First-Principles Calculations of Hydrogen Solution and Diffusion in 3C-SiC Grain Boundaries

Yanan Cui [1], Jingjing Sun [1,*], Meng Li [1] and Bingsheng Li [2]

1 Zhengzhou Key Laboratory of Low-Dimensional Quantum Materials and Devices, College of Physics and Optoelectronic Engineering, Zhongyuan University of Technology, Zhengzhou 450007, China; 2022003065@zut.edu.cn (Y.C.); limeng@zut.edu.cn (M.L.)

2 State Key Laboratory of Environmental Friendly Energy Materials, Southwest University of Science and Technology, Mianyang 621010, China; libingshengmvp@163.com

* Correspondence: sj1993@mail.ustc.edu.cn; Tel.: +86-187-9027-5932

Abstract: First-principles calculations were employed to study the solution and diffusion properties of hydrogen (H) at the Si-rich and C-rich $\Sigma3(111)[1\bar{1}0]$ ($\Sigma3_{Si}$ and $\Sigma3_C$) and $\Sigma9(221)[1\bar{1}0]$ ($\Sigma9$) grain boundaries (GBs) in 3C-SiC. We constructed GBs of varying sizes and calculated their formation energies and excess volumes to identify the stability of GBs. The $\Sigma9$ GB is more stable and has a relatively open structure compared with the $\Sigma3$ GB. The solution energies of H at the $\Sigma3_{Si}$, $\Sigma3_C$ and $\Sigma9$ GBs are significantly reduced to 1.46, 2.30 and 1.47 eV, respectively. These values are much lower than that in the bulk. The negative segregation energies indicate that H is more likely to reside at the GBs rather than in the bulk. The diffusion energy barrier of H in the $\Sigma3_C$ GB is as high as 1.27 eV, whereas in the $\Sigma3_{Si}$ GB and $\Sigma9$ GB, the barriers are as low as 0.42 eV and 0.28 eV, respectively. These results suggest that H migration will be suppressed in the $\Sigma3_C$ GB but promoted in the $\Sigma3_{Si}$ and $\Sigma9$ GBs. The differences in H diffusion behavior among these three GBs may be attributed to the relatively more open structures of the $\Sigma3_{Si}$ and $\Sigma9$ GBs compared with the $\Sigma3_C$ GB. These results are essential for understanding the diffusion mechanism of H and its retention behavior in SiC.

Keywords: 3C-SiC; grain boundaries; hydrogen; diffusion

1. Introduction

Silicon carbide (SiC) ceramic materials exhibit many excellent properties, including high breakdown voltage, a wide bandgap, and resistance to high temperature, high pressure, corrosion, and radiation damage. Due to their superior properties, 4H-SiC and 6H-SiC are widely utilized in high-power electronic devices [1], while 3C-SiC and SiC composites—predominantly consisting of 3C-SiC—are regarded as promising candidates for structural materials in high-temperature nuclear fission reactors and future fusion reactors [2,3]. In high-temperature gas-cooled reactors, SiC is used as the fission products' barrier for three-layer-coated isotropic spherical fuel particles [4]. For nuclear fusion applications, incorporating a thin SiC layer on tungsten or eurofer is expected to enhance the control of the tritium inventory [5–7]. Additionally, using SiC inserts in tritium blankets should improve corrosion resistance [7]. In these applications, SiC is radiated by high-energy particles in the reactors, including high-energy neutrons and hydrogen (H) isotopes, which produce a large number of point defects in its interior and induce amorphization in SiC [8]. Furthermore, transmutation results in the formation of solid elements such as magnesium and boron, as well as gaseous elements including H [9,10]. Irradiation-induced defects and grain boundaries (GBs) are generally the trapping source for H in solids [11,12]. H exhibits a series of behaviors inside SiC, such as diffusion, clusters, and interactions with various defects, particularly GBs [13,14]. These behaviors lead to the formation of H bubbles and volume expansion in SiC, thereby affecting material properties and reducing service life [15]. On the other hand, the diffusion of the H isotope tritium in SiC can also affect the self-sustaining tritium inventory and tritium safety in fusion reactors [16].

The experimental and simulation results of past studies show that GBs play a crucial role in the behaviors of impurities in SiC [17–20]. The influence of different GBs in SiC on the diffusion behavior of impurity atoms is distinct, which may be related to the GB structures [21–23]. The coincident site lattice (CSL)-related GBs (e.g., $\Sigma 3^n$) occupy the majority of GBs in SiC [24,25]. So far, most studies have focused on the behavior of H in SiC bulks, but the interaction mechanism between H and SiC GB is still unclear. The diffusion barrier of H in 3C-SiC bulk was calculated to be 0.5 eV by density functional theory, and the diffusion barriers of H in the presence of silicon (Si) and carbon (C) vacancies are, respectively, as high as 1.09 and 1.71 eV [26,27]. Esteban et al. and Tam et al. measured the diffusion energy barriers of deuterium and tritium in SiC by experimental observations, obtaining values of about 1.12 eV [16,28]. However, the tritium diffusion energy barrier was reported to be 2.82 eV by Causey et al. [29]. It can be found that the values for H exhibit large discrepancies. The difference may result from the different research techniques and the trapping effect of defects for H in SiC. Various defects, including vacancies, interstitial atoms, and GBs, can significantly influence H diffusion within SiC. Among these, GBs represent important intrinsic defects in SiC and may play a crucial role in facilitating the diffusion of H atoms and the formation of H bubbles. Therefore, studying the solution and diffusion behaviors of H in 3C-SiC GBs can help us better understand the experimental results and the material properties.

In this paper, first-principles calculations were used to study the diffusion behaviors of H at the three GBs: Si-rich and C-rich $\Sigma3(111)[1\bar{1}0]$, and $\Sigma9(221)[1\bar{1}0]$ in 3C-SiC. Firstly, we constructed the structures of the three GBs with different C-Si layers and obtained the GB energy. The influence of different C-Si layers on the H solution was investigated in the GB supercells. Then, the solution and segregation energies of H at the GBs were studied. Finally, the diffusion energy barriers of H along the GBs were calculated to further understand the diffusion mechanism.

2. Computational Details

The first-principles calculations based on density functional theory were performed using the Vienna ab initio simulation software package (VASP) [30,31]. The interaction between ions and electrons was expressed by the projected augmented wave potential (PAW) method. The exchange correlation function was described with the plane wave method (PBE) in generalized gradient approximation (GGA) [32,33]. The maximum number of steps of ion relaxation in the calculation was 300. The plane wave truncation energy was 520 eV, and the convergence force on each atom was 0.01 eV/Å. The equilibrium lattice constant for 3C-SiC was optimized to be 4.38 Å. The diffusion properties were calculated using the climbing image nudged elastic band method (CL-NEB) [34]. The thermodynamic stability of GBs can be evaluated by the GB energy γ_{GB}, which is the excess energy of a system containing GB per unit area [35]. The calculation formula for γ_{GB} is

$$\gamma_{GB} = \frac{E_{GB} - NE_{SiC}}{2S} \tag{1}$$

where E_{GB} is the total energy of the GB supercell, and E_{SiC} is the total energy per formula unit of 3C-SiC. N is the number of SiC formula units appearing in the GB supercell, and

S is the cross section area of GB plane. In addition, GB characteristic is always associated with excess volume ΔV. ΔV is the excessive volume of a system containing GB per unit area [36]. ΔV is calculated by

$$\Delta V = \frac{V_{GB} - N * V_{SiC}}{2S} \tag{2}$$

where V_{GB} is the volume of the GB supercell, and V_{SiC} is the volume of C–Si atomic pair in the bulk. The solution energy E_H^{sol} is the energy required to embed a H atom into SiC bulk or GB, which is given as

$$E_H^{sol} = E_{GB/Bulk+H} - E_{GB/Bulk} - \frac{1}{2}H_2 \tag{3}$$

where $E_{GB/Bulk+H}$ represents the total energy of the GB supercell or the 3C-SiC lattice with an interstitial H, $E_{GB/Bulk}$ represents the total energy of the GB supercell or the 3C-SiC lattice without H, and H_2 refers to the energy of a H molecule. The calculation formula for the segregation energy E_H^{Seg} of H at the GB site is as follows:

$$E_H^{Seg} = E_{GB+H}^{sol} - E_{Bulk+H}^{sol} \tag{4}$$

where E_{GB+H}^{sol} represents the total energy of the GB supercell containing H, and E_{Bulk+H}^{sol} represents the solution energy of the H at the most stable site in the perfect SiC bulk.

3. Results and Discussion

3.1. Hydrogen in Perfect 3C-SiC

The solution and diffusion behaviors of H in a perfect 3C-SiC bulk were investigated as a reference. The supercell of 3C-SiC contains 108 C and 108 Si atoms. There are five initial interstitial sites: the tetrahedral center site of four Si atoms (TSi), the tetrahedral center site of four C atoms (TC), the hexagonal site neighboring three C and three Si atoms (Hex), the bond center site of C and Si atoms (BC), and the site along the anti-bonding direction of C-Si bonds (AB_C) [37,38]. H at the Hex site is not stable, which moves to the TSi site. The calculated solution energies for H at the other four stable interstitial sites are shown in Table 1. The solution energies for H at the TC, TSi, and BC sites are 3.61, 2.93, and 2.60 eV, respectively, much higher than that of H at the AB_C site (2.48 eV). Hence, the solution energies follow the following order: TC > TSi > BC > AB_C, where AB_C is the most stable interstitial site for H. Our results are consistent with the previous calculation results [39–41]. The diffusion path of H in the perfect 3C-SiC bulk and the corresponding diffusion energy barriers of H are shown in Figure 1b,c. The energy barrier for a H atom jumping from an AB_C position on the C atom to the equivalent position AB'_C on the nearest neighboring C atom is 0.50 eV. The H can also rotate around a C atom between AB'_C and AB''_C, and the barrier is just 0.10 eV. It can be known that the migration energy barrier of H in the perfect 3C-SiC bulk is 0.50 eV, the same as the result of Kaukonen et al. [27].

Table 1. The solution energies of H at five different interstitial sites in 3C-SiC bulk. Units are eV.

Site	TC	TSi	Hex	BC	AB_C
This work	3.61	2.93	TSi	2.60	2.48
Ref. [39]	3.05	2.61	TSi	2.41	2.05
Ref. [40]	3.36	2.93	TSi	2.57	2.41

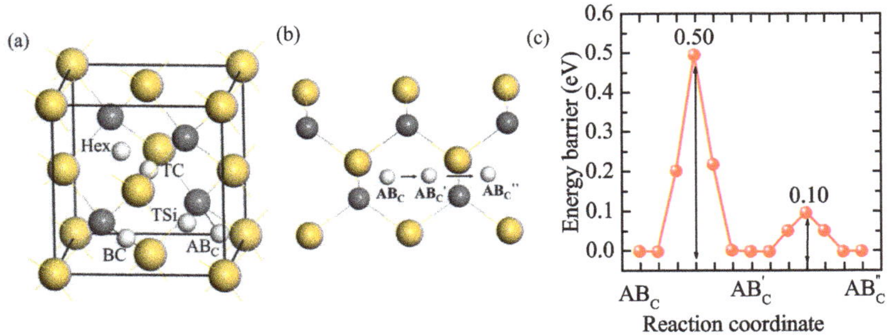

Figure 1. The five initial sites of H in SiC bulk (**a**), the diffusion path (**b**), and diffusion energy barrier (**c**) of H in bulk 3C-SiC. Gray, yellow, and white spheres represent C, Si, and H atoms, respectively.

3.2. The Structures of Different Grain Boundaries

Three GBs, i.e., Si-rich and C-rich $\Sigma3(111)[1\bar{1}0]$ ($\Sigma3_{Si}$ and $\Sigma3_C$) and $\Sigma9(221)[1\bar{1}0]$ ($\Sigma9$), were studied and discussed, and were constructed using the coincidence site lattice model. The structures of $\Sigma3_{Si}$, $\Sigma3_C$, $\Sigma9$, and the basic structural units in the GBs are shown in Figure 2. The supercells of $\Sigma3_{Si}$ and $\Sigma3_C$ contain six C-Si atomic layers, and fifteen C-Si atomic layers are in the $\Sigma9$ supercell. The structural units of $\Sigma3_{Si}$ and $\Sigma3_C$ GBs have a high similarity, all of which contain six atom rings [42,43]. $\Sigma9$ GB is composed of two different polyhedrons with seven-atom rings and five-atom rings, respectively. The excess volumes of the GBs were calculated to be 0.27, 0.07, and 0.07 Å for $\Sigma9$, $\Sigma3_C$, and $\Sigma3_{Si}$, respectively. The excess volume of $\Sigma9$ is much higher than that of $\Sigma3$, indicating that $\Sigma9$ has a much more open structure than $\Sigma3$. Since both the supercells of $\Sigma3_{Si}$ and $\Sigma3_C$ contain a Si-rich GB and a C-rich GB, the excess volumes are the same. The Si-Si bond length in the Si-rich GB is 2.40 Å, and the C-C bond length in the C-rich GB is 1.64 Å. The Si-rich GB is more open than the C-rich GB.

As shown in Figure 2, no vacuum layer was added when constructing the GB structures. To reduce the effects of interactions between periodic images in subsequent calculations, the GB size should be large enough to eliminate interactions between the two GBs, and the system should contain a bulk-like environment in the supercell. The structures of GBs with different C-Si atomic layers were constructed, and the GB energies were calculated and are shown in Table 2. The parameters of $\Sigma3_{Si}$ and $\Sigma3_C$ GBs are the same, because their supercells are essentially the same and the γ_{GB} is the average energy of a Si-rich GB and a C-rich GB in the supercell. The γ_{GB} of $\Sigma3_C$ and $\Sigma3_{Si}$ GB with 6, 8, and 10 C-Si atomic layers are 1.39 J/m^2. The γ_{GB} of $\Sigma9$ GB with 15, 23, and 31 C-Si atomic layers are 1.35 J/m^2. The values of γ_{GB} for the GB systems agree well with the previous calculation results [17]. In addition, for $\Sigma9$ GB, a $3 \times 1 \times 1$ supercell containing 23 C-Si atomic layers was calculated, and other supercells are $2 \times 1 \times 1$, which verifies that in the $[1\bar{1}0]$ direction, the periodic interaction has no effect on the results. As shown in Table 2, it can be found that the γ_{GB} of the three GBs does not change with the increase in atomic layer number. The results suggest that GB sizes are large enough to eliminate interactions between the two GBs in the supercell. To verify that the GBs contain a bulk-like structure, the changes in C-Si atomic layer spacing (Δ_{ij}) in the supercell were calculated. Δ_{ij} is defined as the percentage change in the vertical positions of two subsequent atomic layers (d_i and d_j) with respect to the inter-planar distance (d) in the bulk system, which can be written as $\Delta_{ij} = [(d_j - d_i) - d]/d$. The changes in atomic layer spacing in the [111] direction for $\Sigma3_C$ GB and in the [221] direction for $\Sigma9$ GB are presented in Figure 3 as a function of the layer away from the grain boundary plane. In the other two directions, the changes are neglected. Figure 3a displays the changes in the layer spacing of $\Sigma3_C$ GBs with 6, 8, and 10 C-Si atomic layers.

It can be seen that the changes are close to zero, revealing very small relaxations between the atomic layers. However, in Figure 3b–d, the spacing changes of Σ9 GB with 15, 23, and 31 C-Si atomic layers are different. It can be found that the change in layer spacing near the GB fluctuates greatly and shows a trend of oscillation. The expansion of Δ_{23} is up to about 18%, and the contraction of Δ_{34} is up to about 28%. In the 31 atomic layers of the Σ9 GB, the negligible changes for the layers from 12 to 22 indicate that the atomic relaxation is localized only within 12 layers from the GB. Therefore, the supercell containing 31 layers between the two GBs is large enough to reflect the characteristics of the GB and provides a bulk-like area.

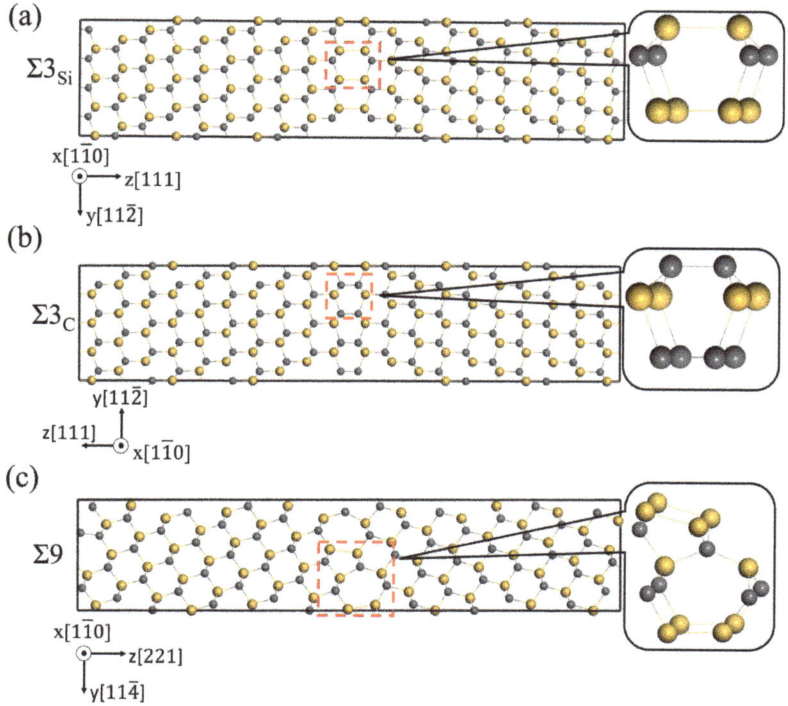

Figure 2. Structures of three GBs, $\Sigma3_{Si}$ (**a**), $\Sigma3_C$ (**b**), and Σ9 (**c**), projected in the $[1\bar{1}0]$ direction. Gray and yellow spheres are used to represent C and Si atoms, respectively.

Table 2. Supercell size, K point, atomic number, and GB energy of $\Sigma3_{Si}$, $\Sigma3_C$, and Σ9 with different C-Si layers.

GB	Layer	L_x (Å)	L_y (Å)	L_z (Å)	K Point	Atoms	γ_{GB} (J/m^2)
$\Sigma3_{Si}$	6	6.19	10.72	30.66	4 × 2×1	192	1.39
	8	6.20	10.72	40.77	4 × 2×1	256	1.39
	10	6.19	10.72	50.89	4 × 2×1	320	1.39
$\Sigma3_C$	6	6.19	10.72	30.66	4 × 2×1	192	1.39
	8	6.20	10.72	40.77	4 × 2×1	256	1.39
	10	6.19	10.72	50.89	4 × 2×1	320	1.39
Σ9	15	6.20	9.34	22.11	4 × 3 × 1	120	1.35
	23	6.20	9.32	33.79	4 × 3 × 1	184	1.35
	23 (3 × 1×1)	9.29	9.32	33.79	4 × 3 × 1	276	1.35
	31	6.20	9.31	45.47	4 × 3 × 1	248	1.35

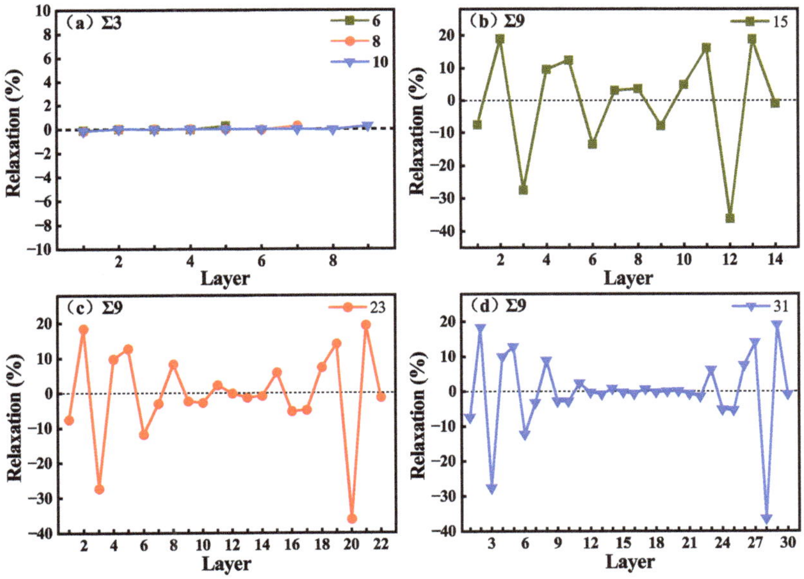

Figure 3. Changes in layer spacing as a function of the number of layers away from the $\Sigma3_C$ GB with 6, 8, and 10 C-Si atomic layers (**a**), and Σ9 GBs with 15 (**b**), 23 (**c**), and 31 (**d**) C-Si atomic layers.

3.3. The Solution Energy of Hydrogen near the Grain Boundaries

The solution behavior of H at the AB_C site on different C-Si layers near the GB in the supercell was calculated, and the solution energy for H at different neighboring locations along the GB is shown in Figure 4, where the black dotted line represents the lowest solution energy of H in the perfect system. Figure 4a is the solution energy of H near the $\Sigma3_C$ GB, and the $\Sigma3_C$ GBs have three different C-Si atomic layers. It can be seen that the three curves of the solution energies show the same trend. The solution energy of H near the 6-layer $\Sigma3_C$ GB is the largest, and the solution energy of H near the 8-layer and 10-layer $\Sigma3_C$ GBs are very close, but only the curve of the solution energies along the 10-layer $\Sigma3_C$ GB shows convergence. Combined with the content of the previous section, it was decided to use 10-layer $\Sigma3_C$ and $\Sigma3_{Si}$ GBs for subsequent research. Figure 4b shows the solution energy of H in the vicinity of Σ9 GB, and the Σ9 GB contains three different C-Si atomic layers and a $3 \times 1 \times 1$ structure of 23 atomic layers. For the 15, 23, and 31-layer GBs, the solution energies of H on the nearest 8, 12, and 16 layers were calculated, respectively. It can be seen that the variation trend and the value of the solution energy of H on the first six layers of the four GBs are almost similar, and the differences become relatively large thereafter. The solution energy begins to converge steadily from the sites of the tenth atomic layer. By comparing the solution energy curves of H near the two different 23-layer GBs, it can be found that they are very close to each other. Hence, the thickness change in the $[1\bar{1}0]$ direction has little effect on the results, and the $2 \times 1 \times 1$ supercell of the 31-layer GB was used for subsequent calculations. In addition, the influence range of Σ9 GB on the solution energy of H is greater than that of the $\Sigma3_C$ GB. This is attributed to the more open structure of the Σ9 GB compared with the $\Sigma3_C$ GB [44]. Overall, the solution energy of H near the GBs is significantly lower than that of H in the bulk-like regions. This suggests that GBs may be a deep trapping region for H atoms.

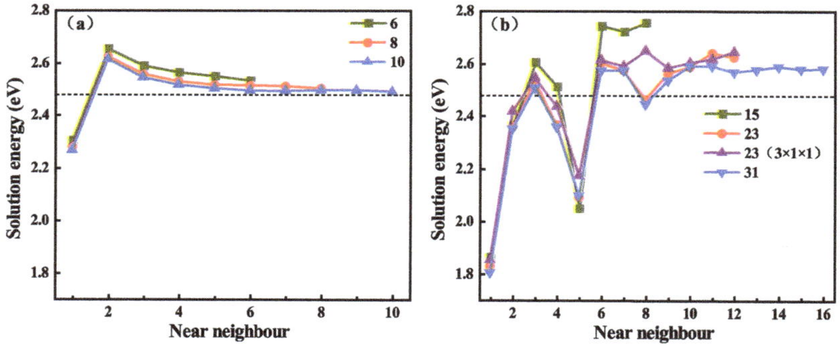

Figure 4. The solution energies of H at different AB_C sites near the $\Sigma3_C$ GBs with 6, 8, and 10 C-Si atomic layers (**a**) and $\Sigma9$ GBs with 15, 23, and 31 C-Si atomic layers (**b**), respectively.

3.4. The Solution and Segregation of Hydrogen at Grain Boundaries

To explore the diffusion characteristics of H in the GB of SiC, it is necessary to understand its solution and segregation properties. $\Sigma3_C$ and $\Sigma3_{Si}$ GBs have similar structures and high symmetry, and the initial interstitial sites for H were selected as similar high symmetry ones. In the $\Sigma9$ GB, eleven possible initial sites were investigated. After structural optimization, the stable positions for H in the three GBs are shown in Figure 5a–c. The solution energies and segregation energies for H were calculated using Equations (3) and (4), and shown in Figure 5d,e, respectively. For the $\Sigma3_{Si}$ GB, the two stable interstitial sites for H are the Si-Si bond center (BC_{Si}) and the triangular prism center (TPC) composed of six Si atoms. The solution energies for H in the position of BC_{Si} and TPC are 1.46 and 1.69 eV, and the segregation energies are −1.02 eV and −0.79 eV, respectively. Hence, the lowest-energy interstitial sites for H at $\Sigma3_{Si}$ GB is BC_{Si}. For the $\Sigma3_C$ GB, the two interstitial sites are the center of the C-C bond (BC_C) and BC_{Si}. The solution energy for H at the BC_C site is 2.30 eV, much lower than that for H at the BC_{Si} site (3.48 eV). The corresponding segregation energies are −0.18 eV and 1.00 eV, respectively. This indicates that H prefers the BC_C site rather than the BC_{Si} site. There are four interstitial sites for H at the $\Sigma9$ GB. They are the position in the mid-vertical plane of the Si-Si bond (MP), the center of six Si atoms (SC), the tetrahedral center of four Si atoms (TC_{Si}), and BC_{Si}. The solution energies for H are 1.47, 1.75, 2.72, and 1.81 eV at the four sites, respectively, and the segregation energies are, respectively, −1.01, −0.73, 0.24, and −0.67 eV. The solution energy for H at MP is much lower than those at the other three sites. MP is the most stable site for H at $\Sigma9$. The solution energies of H in $\Sigma3_{Si}$ and $\Sigma9$ are much lower than that in $\Sigma3_C$, which may due to the more open structure of $\Sigma3_{Si}$ and $\Sigma9$ compared with $\Sigma3_C$. Additionally, the solution energies of H in the three GBs are much lower than that in the perfect SiC bulk, and the segregation energies are negative. The results indicate that H is more likely to segregate in the three GBs rather than in the bulk. Our results here are in agreement with that of Meng et al. [18].

To further understand the segregation behavior of H in SiC GBs, the charge density of the above different configurations were calculated here [45]. The charge density maps of $\Sigma3_{Si}$ and $\Sigma9$ GBs without H are shown in Figure 6, where a, b and e, f represent the charge density maps of two different atomic layers without interstitial H in the $[1\bar{1}0]$ direction. The charge densities of H at the sites BC_{Si} and TPC in the $\Sigma3_{Si}$ GB, and MP and SC in the $\Sigma9$ GB are also presented in Figure 6 for comparison. The charge density between H and Si atoms in the BC_{Si} and MP configurations is denser than that in the TPC and SC configurations in $\Sigma3_{Si}$ and $\Sigma9$ GBs, respectively. H at MP and BC_{Si} sites has a stronger bond with its surrounding Si atoms, while H at TPC and SC sites are in a relatively isolated state and has a weak interaction with its surrounding atoms. Figure 7 shows the partial density of states (PDOSs) of H in the MP position of the Σ9 GB. The s orbital of H and the s and p orbitals of the closest Si are shown in the figure. It can be found that the s orbitals of H and Si have a hybridization peak near −11 eV, which indicates that there is a chemical bond formation between them. Therefore, the reason for the low solution energy of H at the MP and BC_{Si} sites may be that the H atom is more active and easy to form chemical bonds with its surrounding atoms, which is similar to the behavior of oxygen, magnesium, and other elements in SiC GBs [17,18].

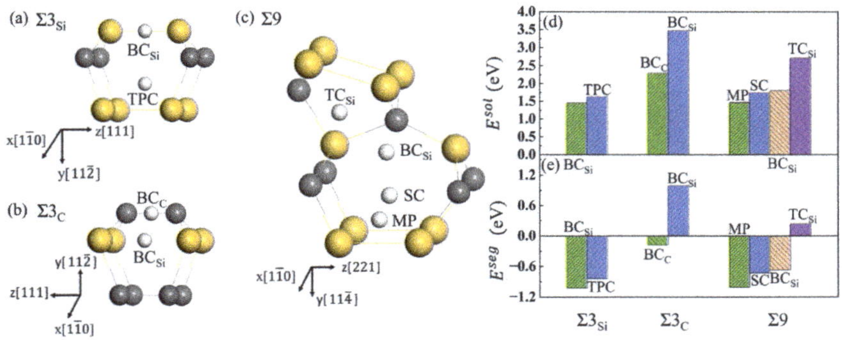

Figure 5. The interstitial sites for H at $\Sigma3_{Si}$ GB (**a**), $\Sigma3_C$ GB (**b**), and Σ9 GB (**c**), and the solution energies (**d**) and segregation energies (**e**) of H in different interstitial positions along the three GBs. BC_{Si} is the position of the Si-Si bond center, and BC_C is the center of the C-C bond. TPC is the site of the triangular prism center, TC_{Si} is the tetrahedral center of four Si atoms, SC is the center of six silicon atoms, and MP is the position in the mid-vertical plane of the Si-Si bond. Gray, yellow, and white spheres represent C, Si, and H atoms, respectively.

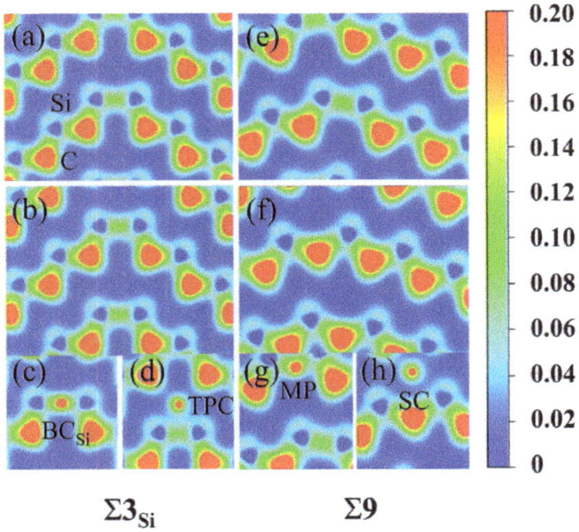

Figure 6. The charge density maps of [1$\bar{1}$0] planes in the $\Sigma3_{Si}$ and Σ9 supercells. (**a**,**b**) are the two different atomic layers without H in the [1$\bar{1}$0] direction of $\Sigma3_{Si}$, while (**e**,**f**) are the two different atomic layers free of H in the [1$\bar{1}$0] direction of Σ9. The charge density maps of H at the BC_{Si} (**c**), TPC (**d**) sites in $\Sigma3_{Si}$ GB and the MP (**g**) and SC (**h**) sites in Σ9 GB are shown. The unit is e/bohr^3.

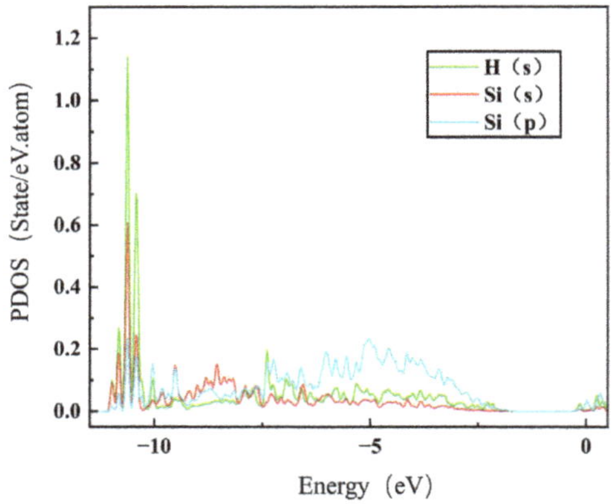

Figure 7. The PDOS for the hybridization between Si and H at the MP site of Σ9.

3.5. The Diffusion of Hydrogen in Grain Boundaries

In the above section, the most stable interstitial sites for H in GBs have been obtained. The diffusion behavior of H along the 3C-SiC GBs was investigated using the CL-NEB method, and the optimal diffusion paths for H in the two directions of GBs were obtained. Figure 8 displays the diffusion paths of H along the two directions at the GB interface, and the corresponding migration energy barriers are shown in Figure 9.

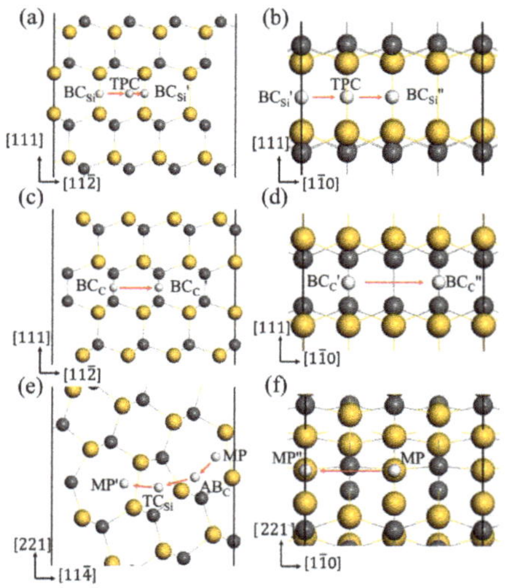

Figure 8. The optimal diffusion paths for H in the directions [11$\bar{2}$] (**a**), [1$\bar{1}$0] (**b**) of $\Sigma3_{Si}$ GB, [11$\bar{2}$] (**c**), [1$\bar{1}$0] (**d**) of $\Sigma3_C$ GB, [11$\bar{4}$] (**e**), [1$\bar{1}$0] (**f**) of Σ9 GB.

For the $\Sigma3_{Si}$ GB, BC_{Si} is the most stable site for H. BC'_{Si} and BC''_{Si} in Figure 8a,b are the equivalent sites of BC_{Si}. In the [11$\bar{2}$] and [1$\bar{1}$0] directions, H prefers to migrate between the two most stable sites through the metastable site TPC. In the path BC_{Si}→TPC→BC'_{Si}, H needs to overcome the energy barrier of 0.42 eV to reach the TPC site, and then overcome the energy barrier of 0.25 eV to reach the BC'_{Si} site in Figure 9a. Therefore, the diffusion barrier of H in the [11$\bar{2}$] direction of $\Sigma3_{Si}$ GB is 0.42 eV. In the [1$\bar{1}$0] direction, the diffusion path for H atom is BC'_{Si}→TPC→BC''_{Si}, and the energy barrier is 0.43 eV in Figure 9b. Hence, the diffusion activation energies of H in the two directions of $\Sigma3_{Si}$ GB are lower than that in the bulk region (0.50 eV), indicating that the $\Sigma3_{Si}$ GB may accelerate H migrating in 3C-SiC.

When H diffuses along the $\Sigma3_C$ GB, the migration paths for H along the $[11\bar{2}]$ and $[1\bar{1}0]$ directions are $BC_C \rightarrow BC'_C$ and $BC'_C \rightarrow BC''_C$ in Figure 8c,d, respectively. The activation energy barrier for H migrating between the BC_C and BC'_C sites is as high as 1.27 eV in Figure 9c, and H needs to overcome the energy barrier of 0.99 eV to reach the BC''_C position from the BC'_C site in Figure 9d. The diffusion energy barriers of H in the two directions along the $\Sigma3_C$ GB are much larger than that of H in perfect SiC bulk. This suggests that H diffusion in the $\Sigma3_C$ GB plane is more difficult than in the bulk.

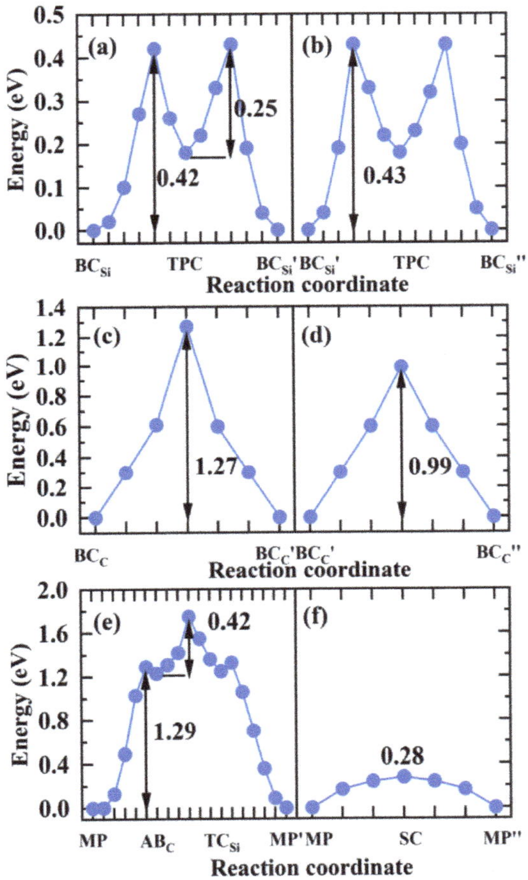

Figure 9. The diffusion energy profile of H atom migrating in the directions $[11\bar{2}]$ (**a**), $[1\bar{1}0]$ (**b**) of $\Sigma3_{Si}$ GB, $[11\bar{2}]$ (**c**), $[1\bar{1}0]$ (**d**) of $\Sigma3_C$ GB, $[11\bar{4}]$ (**e**), $[1\bar{1}0]$ (**f**) of $\Sigma9$ GB.

For the $\Sigma9$ GB, we investigated H migrating between the most stable interstitial sites MP and its equivalent sites MP′ and MP″. The diffusion path for H along the $[11\bar{4}]$ direction is $MP \rightarrow AB_C \rightarrow TC_{Si} \rightarrow MP'$, as shown in Figure 8e. Starting from the MP site, H needs to overcome the energy barrier of 1.29 eV to reach the AB_C site, then it must surmount the energy barrier of 0.42 eV to reach the TC_{Si} site in the 5-atom ring, ultimately reaching the stable site of MP′. Therefore, the diffusion barrier of H in the $[11\bar{4}]$ direction of $\Sigma9$ GB is 1.29 eV, much larger than 0.5 eV. The diffusion path for H in the $[1\bar{1}0]$ direction is shown in Figure 8f, and the diffusion energy barrier is shown in Figure 9f. H hops between MP to MP″ and proceeds via the metastable site SC. The diffusion barrier is 0.28 eV, much lower than 0.5 eV. The results show that the diffusion of H in the two directions of $\Sigma9$ GB is anisotropic. The diffusion of H along the $[11\bar{4}]$ direction is difficult, while the diffusion of H in the $[1\bar{1}0]$ direction is easy. The MP→MP″ along the $[1\bar{1}0]$ direction of the Σ9 GB is likely to serve as a fast diffusion path for H. This is the same as the diffusion behavior of helium in the Σ9 GB of 3C-SiC [46]. This phenomenon may be attributed to the diffusion path of H within an open structure containing seven-atom rings along the $[1\bar{1}0]$ direction. While H diffuses along the $[11\bar{4}]$ direction, it needs to pass the region with high atomic density, so the diffusion energy barrier is relatively high. This is similar to the diffusion phenomenon of H in the Σ9 GB of γ-Fe obtained by He et al. [47].

4. Conclusions

The solution and diffusion properties of H in the three grain boundaries (GBs), Si-rich and C-rich $\Sigma3(111)[1\bar{1}0]$ ($\Sigma3_{Si}$ and $\Sigma3_C$) and $\Sigma9(221)[1\bar{1}0]$ (Σ9) of 3C-SiC, were studied using first-principles calculations. The dimensions of the constructed GB supercells are sufficiently large to ensure the accuracy of the investigations. The GB energy of Σ9 was found to be 1.35 eV, slightly smaller than that of Σ3 (1.39 eV). The solution energies of H in the region near the GBs are significantly lower than that in the bulk. The negative segregation energies indicate that H is more likely to segregate in the three GBs rather than in the bulk. H in the GBs exhibits a greater tendency to form stable chemical bonds with its nearest neighboring atoms. The diffusion energy barrier of H in the $\Sigma3_C$ GB is as high as 1.27 eV, whereas in the $\Sigma3_{Si}$ GB and Σ9 GB, the barriers are as low as 0.42 and 0.28 eV, respectively. These results suggest that the migration of H in the $\Sigma3_C$ GB will be suppressed, but promoted in the $\Sigma3_{Si}$ GB and Σ9 GB. The differences in H diffusion behavior among these three GBs may be attributed to the relatively more open structures of the $\Sigma3_{Si}$ and Σ9 GBs compared with the $\Sigma3_C$ GB. Our theoretical findings provide foundational data for understanding the diffusion mechanism of H and its retention behavior at GBs in SiC.

Author Contributions: Conceptualization, Y.C., J.S., M.L. and B.L.; data curation, Y.C. and J.S.; software, J.S. and M.L.; methodology, J.S. and B.L.; formal analysis, Y.C. and J.S.; investigation, Y.C.; resources, J.S. and M.L.; project administration, J.S.; supervision, J.S. and B.L.; validation, J.S.; writing—original draft preparation, Y.C.; writing—review and editing, J.S. All authors have read and agreed to the published version of the manuscript.

Funding: This work was supported by the National Natural Science Foundation of China (No. 12005320), the Sichuan Science Technology Program (No. 2025NSFJQ0044), the funding scheme for young backbone teachers of Zhongyuan University of Technology (No. 2023XQG12), and the Natural Science Foundation Program of Zhongyuan University of Technology (No. K2025ZD004).

Institutional Review Board Statement: Not applicable.

Informed Consent Statement: Not applicable.

Data Availability Statement: The original contributions presented in this study are included in the article. Further inquiries can be directed to the corresponding author.

Conflicts of Interest: The authors declare that they have no known competing financial interests or personal relationships that could have appeared to influence the work reported in this paper.

References

1. Cheng, Q.Y.; Chen, Z.Y.; Hu, S.S.; Raghothamachar, B.; Dudley, M. Analysis of Threading Edge Dislocation Low-Angle Grain Boundary Network Distributions in 4H-SiC Wafers Through Synchrotron X-ray Topography and Ray-Tracing Simulation. *J. Electron. Mater.* **2025**, *54*. [CrossRef]
2. Katoh, Y.; Snead, L.L.; Szlufarska, I.; Weber, W.J. Radiation effects in SiC for nuclear structural applications. *Curr. Opin. Solid State Mater. Sci.* **2012**, *16*, 143–152. [CrossRef]
3. Olander, D. Nuclear fuels–Present and future. *J. Nucl. Mater.* **2009**, *389*, 1–22. [CrossRef]

4. Lee, Y.W.; Kim, Y.M.; Kim, W.K.; Kim, W.J.; Park, J.Y.; Cho, M.S. Considerations for Some Properties of Pyrolytic Carbon Coating layers in HTR Coated Particle Fuels. In Proceedings of the Transactions of the Korean Nuclear Society Spring Meeting, Jeju, Republic of Korea, 10–11 May 2007. Available online: https://inis.iaea.org/records/80bg7-0pr80 (accessed on 27 March 2025).
5. Gavarini, S.; Baillet, J.; Millard-Pinard, N.; Garnier, V.; Peaucelle, C.; Jaurand, X.; Bernard, C.; Rapegno, R.; Cardinal, S.; Dalverny, C.; et al. Effects of helium irradiation on fine grained β-SiC synthesized by spark plasma sintering. *J. Eur. Ceram. Soc.* **2020**, *40*, 1–11. [CrossRef]
6. Wright, G.M.; Durrett, M.G.; Hoover, K.W.; Kesler, L.A.; Whyte, D.G. Silicon Carbide as a tritium permeation barrier in tungsten plasma-facing components. *Curr. Opin. Solid State Mater. Sci.* **2015**, *458*, 272–274. [CrossRef]
7. Soto, C.; García-Rosales, C.; Echeberria, J.; Martínez-Esnaola, J.M.; Hernández, T.; Malo, M.; Platacis, E.; Muktepavela, F. SiC-based sandwich material for Flow Channel Inserts in DCLL blankets: Manufacturing, characterization, corrosion tests. *Fusion. Eng. Des.* **2017**, *124*, 958–963. [CrossRef]
8. Cai, Z.Q.; Yuan, X.W.; Xu, C.; Li, Y.M.; Shao, Z.; Li, W.J.; Xu, J.X.; Zhang, Q.M. Grain boundary effects on chemical disorders and amorphization-induced swelling in 3C-SiC under high-temperature irradiation: From atomic simulation insight. *J. Eur. Ceram. Soc.* **2024**, *44*, 6911–6925. [CrossRef]
9. Koyanagi, T.; Katoh, Y.; Nozawa, T.; Snead, L.L.; Kondo, S.; Henager, C.H., Jr.; Ferraris, M.; Hinoki, T.; Huang, Q. Recent progress in the development of SiC composites for nuclear fusion applications. *J. Nucl. Mater.* **2018**, *511*, 544–555. [CrossRef]
10. Sawan, M.; Katoh, Y.; Snead, L.L. Transmutation of silicon carbide in fusion nuclear environment. *J. Nucl. Mater.* **2013**, *442*, S370–S375. [CrossRef]
11. Valles, G.; Panizo-Laiz, M.; González, C.; Martin-Bragado, I.; González-Arrabal, R.; Gordillo, N.; Iglesias, R.; Guerrero, C.L.; Perlado, J.M.; Rivera, A. Influence of grain boundaries on the radiation-induced defects and hydrogen in nanostructured and coarse-grained tungsten. *Acta Mater.* **2017**, *122*, 277–286. [CrossRef]
12. Panizo-Laiz, M.; Díaz-Rodríguez, P.; Rivera, A.; Valles, G.; Martín-Bragado, I.; Perlado, J.M.; Munnik, F.; González-Arrabal, R. Experimental and computational studies of the influence of grain boundaries and temperature on the radiation-induced damage and hydrogen behavior in tungsten. *Nucl. Fusion* **2019**, *59*, 086055. [CrossRef]
13. Daghbouj, N.; Li, B.S.; Karlik, M.; Declemy, A. 6H-SiC blistering efficiency as a function of the hydrogen implantation fluence. *Appl. Surf. Sci.* **2019**, *466*, 141–150. [CrossRef]
14. Barcz, A.; Kozubal, M.; Jakieła, R.; Ratajczak, J.; Dyczewski, J.; Gołaszewska, K.; Wojciechowski, T.; Celler, D.G. Diffusion and impurity segregation in hydrogen-implanted silicon carbide. *J. Appl. Phys.* **2014**, *155*, 223710. [CrossRef]
15. O'Connell, J.H.; Neethling, J.H. Investigation of radiation damage and hardness of H^- and He^- implanted SiC. *Radiat. Eff. Defects Solids* **2012**, *167*, 299–306. [CrossRef]
16. Esteban, G.A.; Perujo, A.; Legarda, F.; Sedano, L.A.; Riccardi, B. Deuterium transport in SiC_f/SiC composites. *J. Nucl. Mater.* **2002**, *307*, 1430–1435. [CrossRef]
17. Atkinson, C.M.; Guziewski, M.C.; Coleman, S.P.; Nayak, S.K.; Alpay, S.P. First principles analysis of impurities in silicon carbide grain boundaries. *Acta. Mater.* **2021**, *221*, 117421. [CrossRef]
18. Meng, Z.C.; Wang, C.L.; Wang, Y.L.; Liu, Y.W.; Shu, Y.F.; Yang, L. Screening and manipulation by segregation of dopants in grain boundary of Silicon carbide: First-principles calculations. *Ceram. Int.* **2023**, *49*, 32478–32489. [CrossRef]
19. Leng, B.; Ko, H.; Gerczak, T.J.; Deng, J.; Giordani, A.J.; Hunter, J.L., Jr.; Morgan, D.; Szlufarska, I.; Sridharan, K. Effect of carbon ion irradiation on Ag diffusion in SiC. *J. Nucl. Mater.* **2016**, *471*, 220–232. [CrossRef]
20. Wang, Z.Q.; Zhang, L.M.; AlMotasem, A.T.; Li, B.S.; Polcar, T.; Daghbouj, N. Exploring defect behavior in helium-irradiated single-crystal and nanocrystalline 3C-SiC at 800 °C: A synergy of experimental and simulation techniques. *Acta. Mater.* **2024**, *279*, 120281. [CrossRef]
21. Ding, W.; Lu, P.J.; Xu, Q.F.; Zhang, C.T.F.; Tu, R.; Zhang, S. Grain size and grain boundary characteristics on the out-plane thermal conductivity of <111>-oriented CVD 3C–SiC. *Ceram. Int.* **2024**, *50*, 23775–23783. [CrossRef]
22. Shao, M.Y.; Xu, C.; Hu, R.H.; Lang, Z.; Li, P.W.; Wang, Z.X.; Liu, H.P.; Liu, C.M. Damping effect of (110)<001> symmetric tilt grain boundaries on the shock response of SiC. *Surf. Interfaces* **2025**, *59*, 105992. [CrossRef]
23. Wang, L.; Zhang, L.; Yu, W.S. Revisiting the structures and energies of β-SiC <001> symmetric tilt grain boundaries. *J. Mater. Res.* **2024**, *39*, 2166–2175. [CrossRef]
24. Lillo, T.M.; van-Rooyen, I.J.; Wu, Y.Q. Precession electron diffraction for SiC grain boundary characterization in unirradiated TRISO fuel. *Nucl. Eng. Des.* **2016**, *305*, 277–283. [CrossRef]
25. Lillo, T.M.; van-Rooyen, I.J. Influence of SiC grain boundary character on fission product transport in irradiated TRISO fuel. *J. Nucl. Mater.* **2016**, *473*, 83–92. [CrossRef]
26. Wang, W.Y.; Li, C.; Shang, S.L.; Cao, J.Z.; Liu, Z.K.; Wang, Y.; Fang, C. Diffusion of hydrogen isotopes in 3C-SiC in HTR-PM: A first-principles study. *Prog. Nucl. Energ.* **2020**, *119*, 103181. [CrossRef]
27. Kaukonen, M.; Fall, C.; Lento, J. Interstitial H and H_2 in SiC. *Appl. Phys. Lett.* **2003**, *83*, 923–925. [CrossRef]
28. Tam, S.W.; Kopasz, J.P.; Johnson, C.E. Tritium transport and retention in SiC. *J. Nucl. Mater.* **1995**, *219*, 87–92. [CrossRef]

29. Causey, R.A.; Fowler, J.D.; Ravanbakht, C.; Elleman, T.S.; Verghese, K. Hydrogen Diffusion and Solubility in Silicon Carbide. *J. Nucl. Mater.* **1978**, *61*, 221–225. [CrossRef]
30. Kohn, W.; Sham, L.J. Self-Consistent Equations Including Exchange and Correlation Effects. *Phys. Rev.* **1965**, *140*, A1133. [CrossRef]
31. Kresse, G.; Hafner, J. Ab initio molecular dynamics for liquid metals. *Phys. Rev. B* **1993**, *47*, 558. [CrossRef]
32. Kresse, G.; Joubertr, D. From ultrasoft pseudopotentials to the projector augmented-wave method. *Phys. Rev. B* **1999**, *59*, 1758. [CrossRef]
33. Perdew, J.P.; Zunger, A. Self-interaction correction to density-functional approximations for many-electron systems. *Phys. Rev. B* **1981**, *23*, 5048. [CrossRef]
34. Henkelman, G.; Uberuaga, B.P.; Jónsson, H. A climbing image nudged elastic band method for finding saddle points and minimum energy paths. *J. Chem. Phys.* **2000**, *113*, 9901–9904. [CrossRef]
35. Rohrer, G.S. Grain boundary energy anisotropy: A review. *J. Mater. Sci.* **2011**, *46*, 5881–5895. [CrossRef]
36. Uesugi, T.; Higashi, K. First-Principles Calculation of Grain Boundary Excess Volume and Free Volume in Nanocrystalline and Ultrafine-Grained Aluminum. *Mater. Trans.* **2013**, *54*, 1597–1604. [CrossRef]
37. Roberson, M.A.; Estreicher, S.K. Interstitial hydrogen in cubic and hexagonal SiC. *Phys. Rev. B* **1991**, *44*, 10578. [CrossRef]
38. Aradi, B.; Gali, A.; Deák, P.; Lowther, J.E.; Son, N.T.; Janzén, E.; Choyke, W.J. Ab initio density-functional supercell calculations of hydrogen defects in cubic SiC. *Phys. Rev. B* **2001**, *63*, 245202. [CrossRef]
39. Sun, J.; You, Y.W.; Hou, J.; Li, X.Y.; Li, B.S.; Liu, C.S.; Wang, Z.G. The effect of irradiation-induced point defects on energetics and kinetics of hydrogen in 3C-SiC in a fusion environment. *Nucl. Fusion* **2017**, *57*, 066031. [CrossRef]
40. Zhao, S.Q.; Chen, C.Y.; Ran, G. Effect of H and Pd atoms on the migration of He atoms in 3C–SiC. *Prog. Nucl. Energ.* **2023**, *165*, 104896. [CrossRef]
41. Zhang, L.; Zhang, Y.; Lu, G.H. First-principles investigation of site preference and bonding properties of neutral H in 3C–SiC. *Nucl. Instrum. Methods Phys. Res. Sect. B* **2009**, *267*, 3087–3089. [CrossRef]
42. Rabone, J.; López-Honorato, E. Density functional theory metadynamics of silver, caesium and palladium diffusion at β-SiC grain boundaries. *J. Nucl. Mater.* **2015**, *458*, 56–63. [CrossRef]
43. Möller, H.J. <011> tilt boundaries in the diamond cubic lattice. *Philos. Mag. A* **1981**, *43*, 1045–1055. [CrossRef]
44. Liu, Z.J.; Zhang, Y.G.; Li, X.Y.; Xu, Y.C.; Wu, X.B.; Liu, C.S.; Kong, X.S.; Yao, C.F.; Wang, Z.G. Investigation of the dissolution and diffusion properties of interstitial oxygen at grain boundaries in body-centered-cubic iron by the first-principles study. *RSC. Adv.* **2021**, *11*, 8643–8653. [CrossRef]
45. Krawczuk, A.; Macchi, P. Charge density analysis for crystal engineering. *BMC Chem.* **2014**, *8*, 68. [CrossRef] [PubMed]
46. Sun, J.; You, Y.W.; Wu, X.B.; Song, H.Y.; Li, B.S.; Liu, C.S.; Krsjak, V. Segregation and diffusion behaviours of helium at grain boundaries in silicon carbide ceramics: First-principles calculations and experimental investigations. *J. Eur. Ceram. Soc.* **2022**, *42*, 4066–4075. [CrossRef]
47. He, Y.; Su, Y.J.; Yu, H.B.; Chen, C.F. First-principles study of hydrogen trapping and diffusion at grain boundaries in γ-Fe. *Int. J. Hydrogen Energ.* **2021**, *46*, 7589–7600. [CrossRef]

materials

MDPI

Article

Preparation and Characterization of Graphene-Nanosheet-Reinforced Ni-17Mo Alloy Composites for Advanced Nuclear Reactor Applications

Xiaoxin Ge [1], Yanxin Jiang [1,2], Xu Yu [1], Guopeng Zhang [1], Yunjia Shi [1], Bin Cai [1], Qing Peng [3] and Hai Huang [1,*]

[1] Key Laboratory of Material Physics, Ministry of Education, School of Physics, Zhengzhou University, Zhengzhou 450001, China
[2] International Joint Laboratory for Integrated Circuits Design and Application, Ministry of Education, School of Physics, Zhengzhou University, Zhengzhou 450001, China
[3] State Key Laboratory of Nonlinear Mechanics, Institute of Mechanics, Chinese Academy of Sciences, Beijing 100190, China
* Correspondence: huanghai@zzu.edu.cn

Abstract: Molten salt reactors (MSRs) offer advantages such as enhanced safety, reduced nuclear waste, and cost effectiveness. However, the corrosive nature of fluoride-based molten salts challenges the longevity of structural materials. Ni-based alloys, like Hastelloy N, have shown resistance to fluoride salt corrosion but suffer from issues like helium embrittlement caused by neutron irradiation. To address these concerns, the incorporation of graphene (Gr) into Ni-based alloys is being explored. Gr's superior mechanical properties and irradiation tolerance make it a promising reinforcement material. In this study, a Ni-17Mo alloy, a simplified model of Hastelloy N, was combined with reduced graphene oxide (RGO) using powder metallurgy. The effects of milling time and sintering temperature on the microstructure and mechanical properties were systematically studied. The results indicated that optimal sintering at 1100 °C enhanced tensile strength and ductility. Additionally, RGO incorporation improved the alloy's strength but reduced its elongation. This research highlights the potential of Gr-reinforced Ni-based alloys for advanced MSR applications, offering insights into fabrication techniques and their impact on material properties.

Keywords: graphene–Ni-17Mo alloy composites; powder metallurgy; microstructure; mechanical properties; molten salt reactors

1. Introduction

Molten salt reactors (MSRs) are among the six most advanced Gen-IV nuclear fission reactor technologies, attracting considerable attention due to their inherent safety features, use of anhydrous cooling systems, reduced nuclear waste production, and cost effectiveness [1,2]. Despite these advantages, the highly corrosive nature of fluoride-based molten salt coolants presents significant long-term challenges for structural materials [3]. Conventional reactor materials are inadequate for the extreme operational conditions in MSRs. Ni-based alloys, however, demonstrate exceptional mechanical properties and outstanding resistance to high-temperature corrosion [4]. Leveraging these properties, researchers at Oak Ridge National Laboratory developed the Hastelloy N alloy (i.e., Ni-17 wt.% Mo-7 wt.% Cr), which demonstrated superior resistance to fluoride salt corrosion in the experimental MSRs [5,6]. Nevertheless, its limitations have become increasingly evident during the transition to commercial and industrial applications. During reactor operation, neutron irradiation and the presence of fission products induce microstructural changes in the structural materials, compromising their mechanical integrity and corrosion resistance [7,8]. In addition, Hastelloy N undergoes neutron-induced transmutation reactions, generating helium atoms [9]. These helium atoms tend to accumulate at grain boundaries, leading to reduced ductility and causing helium embrittlement, which severely limits the alloy's service life in reactor cores [9–12]. Consequently, it is imperative to develop new nickel-based alloys with enhanced corrosion resistance, mitigated susceptibility to helium embrittlement, and improved mechanical strength to meet the rigorous demands of future MSR advancements.

The development of irradiation-tolerant materials with self-healing capabilities increasingly focuses on leveraging grain boundaries and heterointerfaces, a growing consensus within the scientific community [13–15]. Graphene (Gr) stands out due to its exceptional electronic and thermal conductivity, coupled with outstanding mechanical properties such as a high Young's modulus (~1 TPa) and intrinsic strength (~130 GPa) [16]. Functionalized Gr is typically synthesized by oxidizing graphite to form graphene oxide (GO) with oxygen-containing functional groups, enhancing its reactivity [17]. The reduction of GO removes these groups, producing reduced graphene oxide (RGO) with vacancy defects [15,18]. As a two-dimensional nanomaterial, Gr and its derivatives show significant promise as a reinforcing component in metals, endowing them with unique functional properties [19]. Recent studies have demonstrated the superior irradiation tolerance of various Gr–metal composites, largely attributed to their ultra-strong and stabilized interfaces [20,21]. For instance, Si et al. [22] reported that Gr/W nanolayers with reduced period thicknesses exhibit remarkable resistance to irradiation, effectively lowering helium bubble density. Similarly, Liu et al. [23] observed that Gr/Al composites outperform unreinforced matrices in reducing irradiation-induced hardening, enhancing elongation, minimizing lattice expansion, and exhibiting unique deformation mechanisms under irradiation. Huang et al. [15] highlighted the ability of the Gr/Ni interface to attract, absorb, and eliminate interstitials, vacancies, and helium atoms or clusters, significantly mitigating residual irradiation defects. These findings suggest that Gr/Ni composites hold strong potential as novel irradiation-tolerant materials for advanced nuclear reactors. Despite this promise, challenges persist in advancing Gr–metal composites as structural materials for MSRs. One major unresolved issue is the interaction between solute elements (e.g., Mo, Cr, and Fe) and Gr's C atoms under high-temperature preparation or irradiation conditions, which may result in metal carbide formation [24]. The binding of these carbides to the Gr surface could alter the interfacial structure, potentially influencing the interfacial defect trapping efficiency.

To attain the desired properties in high-temperature composite fabrication, optimizing the process parameters that govern microstructure control is essential [25]. Powder metallurgy (PM) is widely acknowledged as the preferred method for producing Gr–metal composites [26]. In comparison to casting, PM provides superior control over the microstructure, facilitating a more homogeneous distribution of Gr within the composites [27]. The effectiveness of PM is influenced by several factors, including the duration of ball milling, sintering conditions, and the reinforcement content [28]. Research has demonstrated that the duration of ball milling significantly impacts the microstructural characteristics of the powders [29,30]. Furthermore, sintering parameters—such as temperature, dwell time, and atmosphere—are critical determinants of the composite materials' overall performance [31,32]. Precisely optimizing these parameters, particularly sintering temperature and dwell time, is vital for achieving high density and enhanced mechanical properties in Gr–metal composites. For instance, sintering at excessively low temperatures can result in higher porosity, compromising material performance [33]. Moreover, due to the strong van der Waals forces and hydrophobic nature of Gr, its dispersion in solvents is challenging, leading to agglomeration [34]. In contrast, GO, known for its hydrophilicity, is commonly used in experiments as a substitute [14]. As an example, Zhang et al. [35] employed PM to produce reduced GO (RGO)/Ni composites, showing that, compared to pure Ni, the composites with 0.3 wt.% GO exhibited increases in yield strength, tensile strength, and fracture elongation by 16.3%, 34.3%, and 36%, respectively. Chen et al. [36] fabricated Cu-based composites reinforced with Gr nanosheets at varying concentrations and found that the addition of Gr notably improved the mechanical properties of Cu. However, with increasing Gr content, the strengthening effect initially enhanced and then declined. The maximum elastic modulus and hardness saw improvements of 65% and 75%, respectively, compared to pure Cu. Nonetheless, there remains a paucity of research on the preparation methods for Ni-based alloys incorporating Gr. Furthermore, studies examining the influence of varying Gr content on the microstructural evolution and mechanical properties of Ni-based materials, as well as the associated strengthening mechanisms, are notably lacking.

In this study, the Ni-17Mo alloy was selected as a simplified model for Hastelloy N to explore the fabrication of Gr-nanosheet-reinforced alloy composites using a modified powder metallurgy approach. First, the effects of milling time on the uniformity of powder mixing and particle size were systematically investigated. Then, once the optimal milling time was determined, the powders were sintered at 1000, 1050, and 1100 °C, followed by mechanical property testing of the samples at each temperature. Lastly, the composites were produced by incorporating 0.5 vol.% RGO, and the microstructural characteristics and properties of composites with varying RGO contents were analyzed. Note that this study primarily focuses on the preparation and characterization of the composites by optimizing key processing parameters. A more comprehensive structural analysis and performance assessment, including molten salt corrosion resistance and high-temperature irradiation testing, will be conducted in future research.

2. Experimental Procedures

2.1. Raw Powder Preparation

The Ni powder (purity > 99.6%, Jiangyou Hebao Nanomaterials Co., LTD., Mianyang, China) and Mo powder (Changsha Tianjiu Metal Materials Co., LTD., Changsha, China), both produced via gas atomization, were selected as the raw materials (see Figure 1a,b). The GO, with a purity exceeding 99%, was synthesized using a modified Hummers method [15] and supplied by Nanjing XFnano Material Tech Co. LTD., Nanjing, China (see Figure 1c). A total of 100 g of spherical Ni-17Mo powder was placed into a ball mill tank, and steel balls were added at a ball-to-material mass ratio of 10:1. Argon gas was purged into the tank for 10 min before initiating the milling process. The planetary ball mill was operated at a speed of 300 rpm, with milling times set at 0.5, 1, 2, 4, and 8 h. To prevent excessive heat buildup that could adversely affect the powder properties, the milling process was paused for 5 min after every 0.5 h of operation. The mill's rotation direction was reversed periodically to avoid powder adhering to the tank walls, which could reduce yield. After milling, the suspension was allowed to settle briefly, and the upper liquid was decanted. The residual powder was vacuum-dried at 60 °C for 10 h, resulting in mixed powder with a flake-like morphology. Furthermore, a suspension consisting of GO nanosheets was prepared through ultrasonic dispersion. This involved introducing a measured quantity of GO into deionized water, followed by sonication for 2 h to achieve a uniform suspension with a concentration of 1 g/L. The resulting GO suspension was transferred into a beaker and combined with Ni-17Mo powders that had been previously ball-milled in pure ethanol. The mixture was then subjected to mechanical stirring to ensure complete adsorption of the GO layers onto the flaky Ni-17Mo powder surfaces.

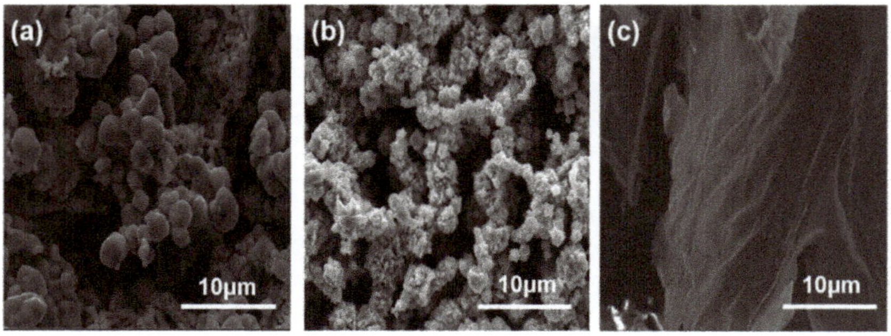

Figure 1. SEM images showing the morphology and particle size distribution of Ni powder (**a**), Mo powder (**b**), and GO powder (**c**) in their unprocessed forms.

2.2. *Synthesis of Alloy and Composite Bulk*

The powder mixture was placed into a graphite mold with a 40 mm diameter. This mold was then positioned in a fast hot-pressing sintering furnace and subjected to sintering under a vacuum pressure of 10^{-2} bar. The sintering was conducted at temperatures of 1000, 1050, and 1100 °C for the Ni-17Mo alloy bulk and at 1100 °C for the 0.5 vol.% RGO/Ni-17Mo alloy composites. Note that this specific Gr concentration was not determined through optimization in this study but was selected based on extensive research indicating its effectiveness in enhancing material properties [35,37,38]. Its adoption aimed to assess the reliability of the proposed fabrication process. During the holding stage, a pressure of 70 MPa was applied for 10 min, with a heating rate of 100 °C/min and a cooling time of approximately 40 min. The final sintered sample had a diameter of 40 mm and a thickness of approximately 6 mm. Detailed information on the rates of pressurization, heating, and cooling throughout the sintering process is provided in Figure 2.

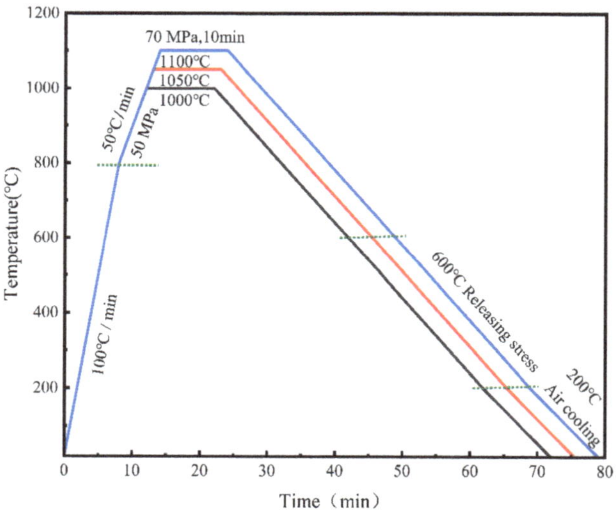

Figure 2. Schematic representation of the heating cycles used during the sintering of Ni-17Mo alloy powder compacts milled for 2 h at various temperatures, along with the processing route for composite powders. Note that dashed lines demarcate each sintering phase to distinguish stage-specific conditions.

2.3. Microstructure Characterization and Property Measurements

The powder's size, morphology, and microstructure were analyzed using two-beam scanning electron microscopy (SEM, Helions G4CX) and X-ray diffraction (XRD, PANalytical Empyrean, Almelo, The Netherlands) with a Cu-$K_{\alpha 1}$ radiation source (λ = 1.5406 Å). SEM analysis was conducted in conjunction with energy-dispersive X-ray spectroscopy (EDS) at an accelerating voltage of 20 kV. XRD patterns were recorded over a 2θ range of 20–90° with a scan rate of 10°/min. The Raman spectra of the original, ultrasonically dispersed, and reduced GO powders were acquired using a Renishaw inVia Raman Microscope, Gloucestershire, UK. The measurements were conducted with a 532 nm He–Ne laser, a spectral resolution of 1 cm^{-1}, and a 100× objective lens (numerical aperture: 0.85) utilizing the line-scan method. Additionally, the mechanical properties of the Ni-17Mo alloy at varying sintering temperatures and 0.5 vol.% RGO/Ni-17Mo alloy composites were evaluated using a universal testing machine (MTS-CMT5105, Eden Prairie, MN, USA) and a digital microhardness tester (HVS-50, Shanghai, China). Microhardness measurements were repeated at least five times with a 4.903 N load and 15 s dwell time. Uniaxial tensile tests were conducted at room temperature with a fixed crosshead speed of 2.0 mm/min.

3. Results and Discussion

3.1. Effect of Milling Time on Powder Mixture

Figure 3 presents the XRD spectrum of Ni-17Mo alloy powder, with milling times ranging from 0 to 8 h. Both the intensity and width of the diffraction peaks decrease and broaden with increasing milling duration. This trend suggests a reduction in the average grain size, accompanied by a rise in lattice distortion and crystal defects. A similar pattern has been reported in previous studies [39]. In addition, milling durations ranging from 0.5 to 8 h did not result in mechanical alloying of the elemental Ni and Mo powders.

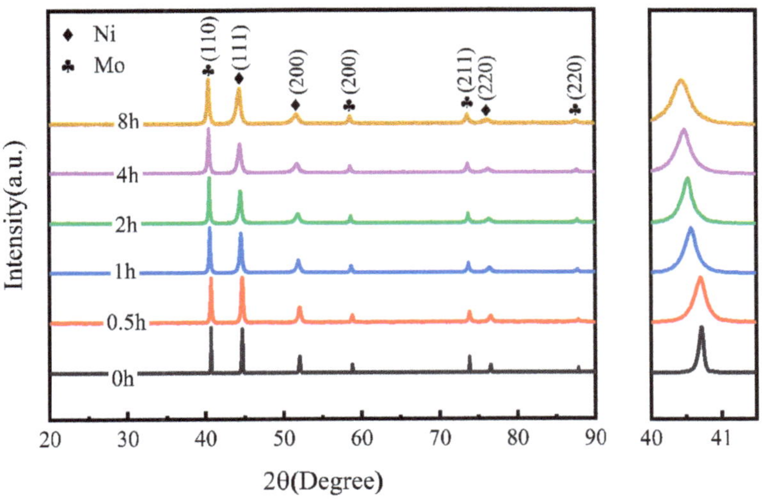

Figure 3. XRD patterns of Ni-17Mo alloy powders milled for different durations, illustrating the evolution of crystallographic phases with milling time.

Figures 4 and 5 illustrate the changes in the size and morphology of the alloy powder particles during ball milling. The average particle size was measured via SEM image analysis using software ImageJ 1.53t (National Institutes of Health, Bethesda, MD, USA), and the resulting size distribution was modeled with a log-normal function, a standard method for analyzing ball-milled powders [40]. Clearly, as milling time increases from 0 to 8 h, the average grain size slightly diminishes from 8.49 to 7.22 μm. Initially, both Ni and Mo powders were nearly spherical in shape. However, after ball milling for various durations, significant alterations in particle morphology and size were observed. Following the milling process, the mixed powder particles exhibited enhanced plastic deformation capabilities. The impact between the grinding balls, powder, and milling jar created a micro-forging effect, leading to the formation of flaky and fragmented powder particles. Through a combination of micro-forging and crushing actions, the particle size was progressively refined. The observed trend in particle size reduction with increasing milling time can be attributed to the dominance of the fracturing mechanism during the process [41]. As the powder is broken down, its particle size decreases. Under identical milling conditions, a smaller particle size typically corresponds to a higher oxygen content due to the increased specific surface area and surface energy, which enhance the powder's ability to adsorb oxygen. Consequently, the powder milled for 8 h likely exhibited the highest oxidation level. Previous studies indicate that excessive oxygen content can result in oxide inclusions and the formation of a continuous oxide film on the powder particles' surface [42]. This oxide layer impedes particle activation and bonding during sintering, leading to reduced density and deteriorating the mechanical properties of the sintered material. No mechanical alloying was observed in the powder milled for 8 h. Therefore, a shorter milling duration is recommended to mitigate the effects of oxidation, as confirmed by the scanned images in Figure 6. Powders milled for 0.5 and 1 h showed uneven mixing, while those milled for 2 h exhibited a more uniform blend, making 2 h the optimal milling time in this study for subsequent sintering.

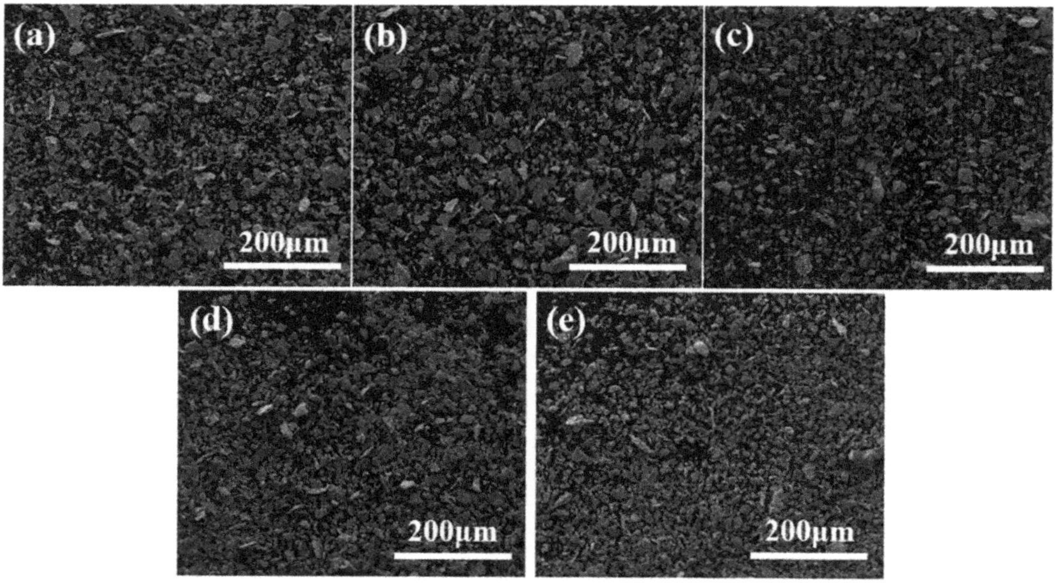

Figure 4. Low-magnification SEM images depicting the morphology and microstructural evolution of Ni-17Mo alloy powder particles after ball milling for 0.5 h (**a**), 1 h (**b**), 2 h (**c**), 4 h (**d**), and 8 h (**e**).

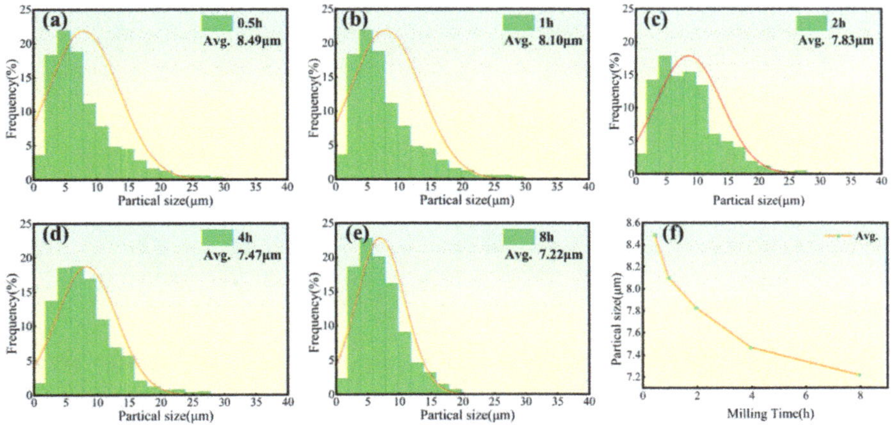

Figure 5. Particle size distribution curves (**a**–**e**) and their average sizes (**f**) for Ni-17Mo alloy powders milled over different durations ranging from 0.5 to 8 h, highlighting the effect of milling time on particle refinement and uniformity.

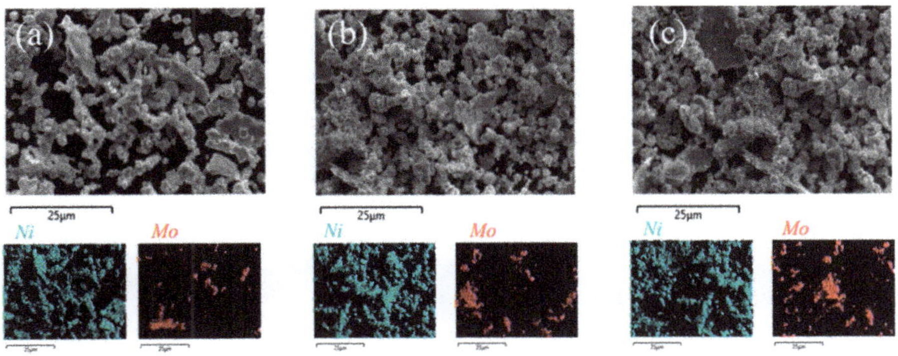

Figure 6. High-magnification SEM images of Ni-17Mo alloy powders milled for 0.5 h (**a**), 1 h (**b**), and 2 h (**c**), along with corresponding EDS elemental maps that illustrate the elemental distribution and microstructural features.

3.2. *Effect of Process Parameters on Properties*

Figure 7 presents the stress–strain curves and relative densities for the sintered Ni-17Mo alloy at various temperatures. Additionally, key mechanical parameters from Figure 7 are tabulated in Table 1 for analysis. The data reveal that the samples sintered at 1000 °C and 1050 °C exhibit similar tensile properties, while the sample sintered at 1100 °C demonstrates superior tensile performance, with a yield strength of 579.5 MPa, a tensile strength of 859.6 MPa, and an elongation of 31.6%. Compared to the lower sintering temperatures, both the tensile strength and elongation are notably improved at 1100 °C. This improvement suggests that a denser structure forms at 1100 °C due to enhanced thermal diffusion. Based on the tensile mechanical properties and relative density data across different sintering temperatures, it can be concluded that the optimal sintering temperature for the Ni-17Mo alloy is 1100 °C.

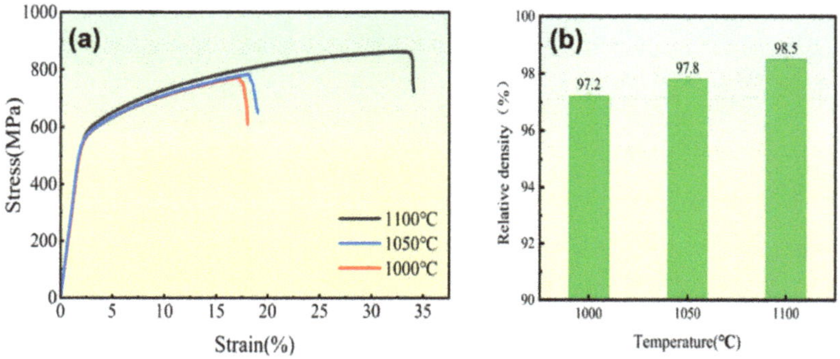

Figure 7. Stress–strain curves (**a**) and relative density measurements (**b**) of sintered Ni-17Mo alloys processed at various sintering temperatures, demonstrating the relationship between sintering conditions and mechanical performance.

Table 1. Statistical summary of mechanical testing results.

Samples	Yield Strength (MPa)	Ultimate Tensile Strength (MPa)	Elongation (%)	Hardness (HV)
1000 °C	544.3 ± 7.4	771.6 ± 5.5	15.9 ± 2.0	259 ± 9
1050 °C	557.2 ± 9.2	784.9 ± 15.2	16.7 ± 3.1	246 ± 7
1100 °C	579.5 ± 4.5	859.6 ± 8.3	31.6 ± 2.5	190 ± 3

3.3. Effect of Adding Gr on Alloy Matrix

Figure 8 displays the SEM image of the RGO/Ni-17Mo composite powder after mixing, along with its corresponding EDS spectrum. The energy dispersive spectroscopy (EDS) analysis reveals that the alloy powder and RGO are uniformly blended, which supports the homogeneity of the composition in the resulting composite material. Additionally, the analysis illustrates the "brick-and-mortar" arrangement of the alloy flake powder and RGO, facilitating the formation of a distinct bionic layered structure within the composite [43–45].

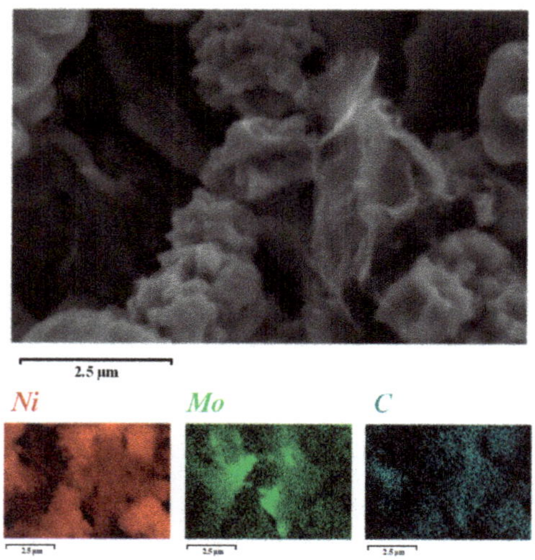

Figure 8. SEM image of 0.5 vol.% RGO/Ni-17Mo alloy composite powder and its corresponding EDS elemental mappings, revealing the dispersion of RGO within the matrix.

Figure 9 presents the Raman spectra of the original, the ultrasonically dispersed, and the reduced GO powders obtained at different stages of the preparation process. Raman spectroscopy allows for monitoring the quality of Gr throughout the preparation process. The blue curve represents the Raman spectrum of the original GO powder, with an I_D/I_G ratio (corresponding to the peak area ratio [46]) of 0.92. The red curve corresponds to the Raman spectrum of composite powder, showing an I_D/I_G ratio of 1.03. This increase indicates further degradation of the Gr structure following the ultrasonic dispersion of GO. This structural damage is linked to a reduction in the diameter of GO sheets caused by the ultrasonic treatment, which leads to a higher volume fraction of Gr with incomplete edges, thus raising the I_D/I_G ratio. The black curve shows the Raman spectrum after thermal reduction at 500 °C, with an I_D/I_G ratio of 0.96. This result demonstrates that high-temperature thermal reduction effectively aids in repairing the Gr structure.

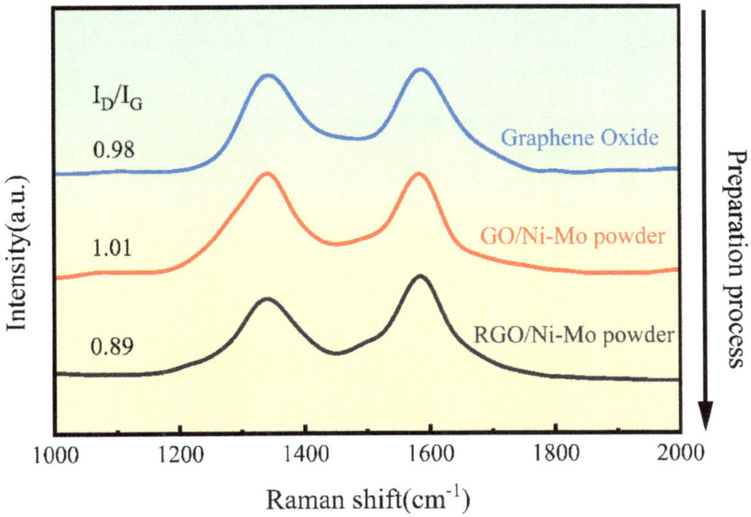

Figure 9. Raman spectra of GO powders at different stages of the preparation process, highlighting changes in defect structures.

Figure 10 presents the room temperature tensile stress–strain curve of RGO/Ni-17Mo composite sintered bodies with varying RGO contents tested at a tensile rate of 10^{-3} mm/s. The sintered Ni-17Mo alloy exhibits a yield strength of 579.5 MPa, a tensile strength of 859.6 MPa, and an elongation at break of 31.6%. With the addition of 0.5 vol.% RGO, the yield and tensile strengths of the composite sintered body increase to 635.9 MPa and 894.3 MPa, respectively, although the elongation at break decreases to 26.4%. This enhancement in strength is attributed to the formation of Mo-rich carbides in the 0.5 vol% RGO/Ni-17Mo alloy composite. As seen in Figure 11, the introduction of RGO not only increases the irregular black contrast structures associated with cavities but also results in an increase in white regions. According to the literature [47], these white areas correspond to Mo-rich carbides. Owing to molybdenum's strong affinity for carbon, the addition of RGO promotes carbide formation. Furthermore, RGO tends to agglomerate and stack, which increases the sample's porosity and hinders matrix densification [48]. Consequently, while carbides enhance the composite's strength and hardness, their presence also induces stress concentrations during tensile deformation, leading to premature fracture and reduced plasticity.

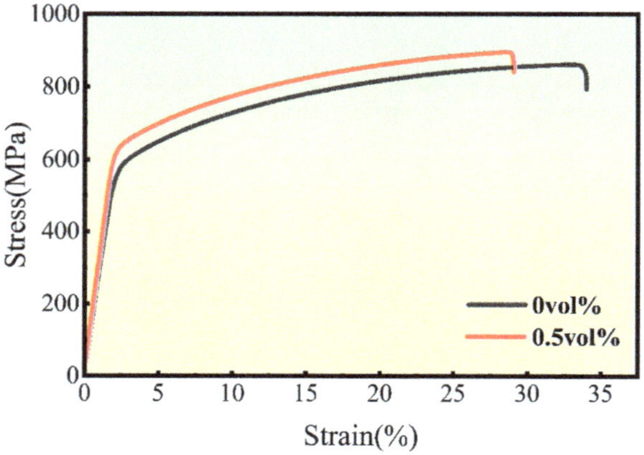

Figure 10. Stress–strain curves for Ni-17Mo alloy composites with different RGO contents, illustrating the influence of RGO on mechanical properties.

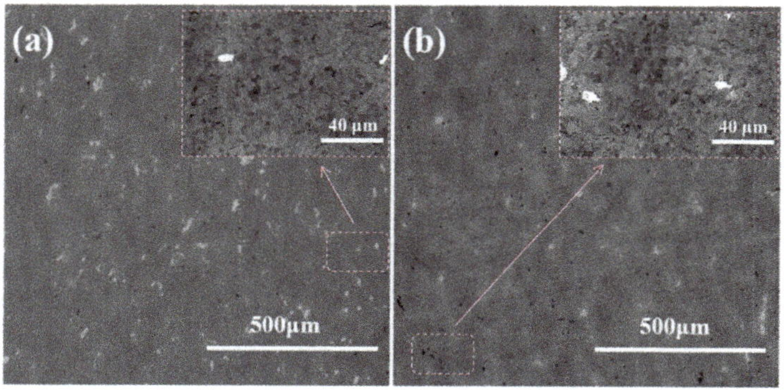

Figure 11. SEM images showing the microstructural variations in Ni-17Mo alloy composites with different RGO contents. (**a**) 0 vol% RGO. (**b**) 0.5 vol% RGO.

4. Conclusions

In summary, this study investigates the fabrication of Gr-nanosheet-reinforced Ni-17Mo alloy composites using a modified powder metallurgy approach. The effects of milling time on powder uniformity and particle size were examined, and optimal milling conditions were determined for subsequent sintering at temperatures of 1000, 1050, and 1100 °C. Mechanical properties of the sintered samples were tested, and composites were created by incorporating 0.5 vol.% RGO. XRD and SEM analyses revealed that milling time affected the alloy powder's grain size and morphology, with the optimum milling duration identified as 2 h to achieve uniform mixing. The sintering temperature significantly influenced the mechanical properties, with samples sintered at 1100 °C showing the best performance, including improved yield strength (579.5 MPa), tensile strength (859.6 MPa), and elongation (31.6%). The addition of 0.5 vol.% RGO to the Ni-17Mo alloy improved strength, reaching 635.9 MPa yield strength and 894.3 MPa tensile strength, though elongation decreased to 26.4%. This enhancement was attributed to the formation of Mo-rich carbides, which strengthened the composite but also led to reduced plasticity due to stress concentration. The Raman spectroscopy results indicated that the Gr structure was slightly degraded during ultrasonic dispersion but was largely repaired by thermal reduction, confirming the importance of processing conditions on the final composite structure. These findings suggest that optimizing milling and sintering parameters is crucial for producing high-performance Gr/Ni-17Mo composites for potential applications in advanced nuclear reactors.

Author Contributions: X.G. conducted the sample preparation and experimental characterizations, helped with data analyses, and interpreted the results. X.Y. and Y.J. provided assistance in experimental characterizations, modifying the manuscript, and interpreting the results. G.Z., Y.S., Q.P. and B.C. provided assistance in the discussions of the data. H.H. designed and supervised the project. All authors have read and agreed to the published version of the manuscript.

Funding: This work was supported by the National Natural Science Foundation of China (Grant No. 12105249), the Key Project for Science and Technology Development of Henan Province (Grant No. 242102230052), and the Henan Province Postdoctoral Science Foundation (Grant No. 202102012).

Institutional Review Board Statement: Not applicable.

Informed Consent Statement: Not applicable.

Data Availability Statement: The original contributions presented in this study are included in the article. Further inquiries can be directed to the corresponding author.

Acknowledgments: We thank the experimental characterization resources provided by the Center of Advanced Analysis & Gene Sequencing, Zhengzhou University.

Conflicts of Interest: The authors declare no conflicts of interest.

References

1. LeBlanc, D. Molten salt reactors: A new beginning for an old idea. *Nucl. Eng. Des.* **2010**, *240*, 1644–1656. [CrossRef]
2. Patel, N.S.; Pavlík, V.; Boča, M. High-Temperature Corrosion Behavior of Superalloys in Molten Salts—A Review. *Crit. Rev. Solid State Mater. Sci.* **2016**, *42*, 83–97. [CrossRef]
3. Guo, S.; Zhang, J.; Wu, W.; Zhou, W. Corrosion in the molten fluoride and chloride salts and materials development for nuclear applications. *Prog. Mater. Sci.* **2018**, *97*, 448–487. [CrossRef]
4. Li, C.; Lei, G.; Liu, J.; Liu, A.; Ren, C.L.; Huang, H. A potential candidate structural material for molten salt reactor: ODS nickel-based alloy. *J. Mater. Sci. Technol.* **2022**, *109*, 129–139. [CrossRef]
5. Vacik, J.; Naramoto, H.; Cervena, J.; Hnatowicz, V.; Peka, I.; Fink, D. Absorption of molten fluoride salts in glassy carbon, pyrographite and Hastelloy B. *J. Nucl. Mater.* **2001**, *289*, 308–314. [CrossRef]
6. Volkov, S.; Omel'chuk, A.; Azhazha, V.; Bakai, A. Corrosion stability of irradiated Hastelloy-type alloys in molten NaF-ZrF4 mixture. *J. New Mater. Electrochem. Syst.* **2006**, *9*, 305–311.
7. Murty, K.L.; Charit, I. Structural materials for Gen-IV nuclear reactors: Challenges and opportunities. *J. Nucl. Mater.* **2008**, *383*, 189–195. [CrossRef]
8. Ye, Y.; Ma, L.; Tang, T.; Liu, T.; Chen, F.; Ge, X.; Jiang, Y.; Yu, X.; Huang, H. Characterization of microstructure and properties of Ti35 alloy and its high-fluence hydrogen irradiation-induced surface exfoliation. *Eur. Phys. J. Plus* **2024**, *139*, 822. [CrossRef]
9. Qiu, S.; Liu, H.; Jiang, M.; Min, S.; Gu, Y.; Wang, Q.; Yang, J.; Li, X.; Chen, Z.; Hou, J. A Brief Review on He Ion Irradiation Research of Steel and Iron-Based Alloys in Nuclear Power Plants. *Acta Metall. Sin. (Eng. Lett.)* **2022**, *36*, 529–551. [CrossRef]
10. Zemła, M.R.; Wróbel, J.S.; Wejrzanowski, T.; Nguyen-Manh, D.; Kurzydłowski, K.J. The helium effect at grain boundaries in Fe-Cr alloys: A first-principles study. *Nucl. Instrum. Methods Phys. Res. Sect. B Beam Interact. Mater. At.* **2017**, *393*, 118–121. [CrossRef]
11. Sharma, S.S.; Parashar, A. Effect of helium on thermal transport properties in single-and bi-crystals of Ni: A study based on molecular dynamics. *J. Phys. D Appl. Phys.* **2023**, *57*, 055304. [CrossRef]
12. Ma, L.; Liu, T.; Cai, B.; Liu, Z.; Zhang, G.; Li, J.; Li, H.; Huang, H. Molecular dynamics studies of primary irradiation damage in α-type Ti35 alloy. *Phys. Status Soldi B* **2023**, *260*, 2200560. [CrossRef]
13. Barr, C.M.; El-Atwani, O.; Kaoumi, D.; Hattar, K. Interplay Between Grain Boundaries and Radiation Damage. *JOM* **2019**, *71*, 1233–1244. [CrossRef]
14. Su, Z.; Jiang, H.; Li, H.; Zhang, Y.; Chen, J.; Zhao, J.; Ma, Y. Recent progress on interfaces in nanomaterials for nuclear radiation resistance. *ChemNanoMat* **2023**, *9*, e202200477. [CrossRef]
15. Huang, H.; Tang, X.; Chen, F.; Liu, J.; Sun, X.; Ji, L. Radiation tolerance of nickel–graphene nanocomposite with disordered graphene. *J. Nucl. Mater.* **2018**, *510*, 1–9. [CrossRef]
16. Nieto, A.; Bisht, A.; Lahiri, D.; Zhang, C.; Agarwal, A. Graphene reinforced metal and ceramic matrix composites: A review. *Int. Mater. Rev.* **2016**, *62*, 241–302. [CrossRef]

17. Razaq, A.; Bibi, F.; Zheng, X.; Papadakis, R.; Jafri, S.H.M.; Li, H. Review on graphene-, graphene oxide-, reduced graphene oxide-based flexible composites: From fabrication to applications. *Materials* **2022**, *15*, 1012. [CrossRef]
18. Sinha, A.; Ranjan, P.; Thakur, A.D. Effect of characterization probes on the properties of graphene oxide and reduced graphene oxide. *Appl. Phys. A* **2021**, *127*, 585. [CrossRef]
19. Chen, D.; Li, J.; Sun, K.; Fan, J. Graphene-reinforced metal matrix composites: Fabrication, properties, and challenges. *Int. J. Adv. Manuf. Technol.* **2023**, *125*, 2925–2965. [CrossRef]
20. Liu, T.; Yuan, X.; Huang, H. Primary irradiation damage in Ni–graphene nanocomposites with pre-existing hydrogen: Insights from atomistic simulations. *Eur. Phys. J. Plus* **2024**, *139*, 22. [CrossRef]
21. Huang, H.; Tang, X.; Xie, K.; Peng, Q. Enhanced self-healing of irradiation defects near a Ni–graphene interface by damaged graphene: Insights from atomistic modeling. *J. Phys. Chem. Solids* **2021**, *151*, 109909. [CrossRef]
22. Si, S.; Li, W.; Zhao, X.; Han, M.; Yue, Y.; Wu, W.; Guo, S.; Zhang, X.; Dai, Z.; Wang, X.; et al. Significant radiation tolerance and moderate reduction in thermal transport of a tungsten nanofilm by inserting monolayer graphene. *Adv. Mater.* **2017**, *29*, 1604623. [CrossRef]
23. Liu, Y.; Zeng, Y.; Guo, Q.; Zhang, J.; Li, Z.; Xiong, D.-B.; Li, X.; Zhang, D. Bulk nanolaminated graphene (reduced graphene oxide)–aluminum composite tolerant of radiation damage. *Acta Mater.* **2020**, *196*, 17–29. [CrossRef]
24. Liu, Y.; Zhang, S.; Yao, Y.; Fan, L.; Wang, J.; Wu, Y.; Jing, L.; Han, P.; Zhang, C. Enhancement of strength-ductility synergy of in situ synthesized graphene/Ni composite via Mo microalloying. *Mater. Sci. Eng. A* **2025**, *919*, 147496. [CrossRef]
25. Gürbüz, M.; Mutuk, T. Effect of process parameters on hardness and microstructure of graphene reinforced titanium composites. *J. Compos. Mater.* **2017**, *52*, 543–551. [CrossRef]
26. Sadhu, K.K.; Mandal, N.; Sahoo, R.R. SiC/graphene reinforced aluminum metal matrix composites prepared by powder metallurgy: A review. *J. Manuf. Process.* **2023**, *91*, 10–43. [CrossRef]
27. Rahimian, M.; Ehsani, N.; Parvin, N.; Baharvandi, H.R. The effect of particle size, sintering temperature and sintering time on the properties of Al–Al_2O_3 composites, made by powder metallurgy. *J. Mater. Process. Tech.* **2009**, *209*, 5387–5393. [CrossRef]
28. Sathish, T.; Saravanan, R.; Kumar, A.; Prakash, C.; Shahazad, M.; Gupta, M.; Senthilkumar, N.; Pandit, B.; Ubaidullah, M.; Smirnov, V.A. Influence of synthesizing parameters on surface qualities of aluminium alloy AA5083/CNT/MoS_2 nanocomposite in powder metallurgy technique. *J. Mater. Res. Technol.* **2023**, *27*, 1611–1629. [CrossRef]
29. Ponhan, K.; Tassenberg, K.; Weston, D.; Nicholls, K.G.M.; Thornton, R. Effect of SiC nanoparticle content and milling time on the microstructural characteristics and properties of Mg-SiC nanocomposites synthesized with powder metallurgy incorporating high-energy ball milling. *Ceram. Int.* **2020**, *46*, 26956–26969. [CrossRef]
30. Salur, E.; Aslan, A.; Kuntoğlu, M.; Acarer, M. Effect of ball milling time on the structural characteristics and mechanical properties of nano-sized Y2O3 particle reinforced aluminum matrix composites produced by powder metallurgy route. *Adv. Powder Technol.* **2021**, *32*, 3826–3844. [CrossRef]
31. Du, S.; Xu, L.; Zhang, X.; Hu, P.; Han, W. Effect of sintering temperature and holding time on the microstructure and mechanical properties of ZrB_2–SiCw composites. *Mater. Chem. Phys.* **2009**, *116*, 76–80. [CrossRef]
32. Kang, P.; Zhao, Q.; Guo, S.; Xue, W.; Liu, H.; Chao, Z.; Jiang, L.; Wu, G. Optimisation of the spark plasma sintering process for high volume fraction SiCp/Al composites by orthogonal experimental design. *Ceram. Int.* **2021**, *47*, 3816–3825. [CrossRef]
33. Hu, C.; Liu, J.; Xu, L.; Yu, L.; Zhu, B. Effect of Sintering Temperature on Properties of Carbon Fiber-Reinforced Titanium Matrix Composites. *ACS Omega* **2022**, *7*, 30087–30092. [CrossRef] [PubMed]
34. Zhan, J.; Lei, Z.; Zhang, Y. Non-covalent interactions of graphene surface: Mechanisms and applications. *Chem* **2022**, *8*, 947–979. [CrossRef]
35. Zhang, X.; Liu, Y.; Liu, X.; Zhang, J.; Chen, X.; Wu, Y.; Jing, L.; Wu, X.; Zhao, M.; Han, P. In-situ grown few-layer graphene reinforced Ni matrix composites with simultaneously enhanced strength and ductility. *Mater. Sci. Eng. A* **2021**, *828*, 142118. [CrossRef]
36. Chen, F.Y.; Ying, J.M.; Wang, Y.F.; Du, S.Y.; Liu, Z.P.; Huang, Q. Effects of graphene content on the microstructure and properties of copper matrix composites. *Carbon* **2016**, *96*, 836–842. [CrossRef]
37. Jiang, R.; Zhou, X.; Liu, Z. Electroless Ni-plated graphene for tensile strength enhancement of copper. *Mater. Sci. Eng. A.* **2017**, *679*, 323–328. [CrossRef]
38. Wang, J.; Li, Z.; Fan, G.; Pan, H.; Chen, Z.; Zhang, D. Reinforcement with graphene nanosheets in aluminum matrix composites. *Scr. Mater.* **2012**, *66*, 594–597. [CrossRef]
39. Li, C.; Zhou, Y.; Xie, Y.; Zhou, D.; Zhang, D. Effects of milling time and sintering temperature on structural evolution, densification behavior and properties of a W-20wt.%Cu alloy. *J. Alloys Compd.* **2018**, *731*, 537–545. [CrossRef]
40. Collins, T.J. ImageJ for microscopy. *Biotechniques* **2007**, *43* (Suppl. S1), 25–30. [CrossRef]
41. Singh, P.; Abhash, A.; Yadav, B.N.; Shafeeq, M.; Singh, I.B.; Mondal, D.P. Effect of milling time on powder characteristics and mechanical performance of Ti4wt%Al alloy. *Powder Technol.* **2019**, *342*, 275–287. [CrossRef]

42. Li, D.; He, H.; Lou, J.; Li, Y.; He, Z.; Chen, Y.; Luo, F. Effect of oxygen contents on predominant sintering mechanism during initial stage of pure titanium powder. *Powder Technol.* **2020**, *361*, 617–623. [CrossRef]
43. Cao, M.; Xiong, D.-B.; Tan, Z.; Ji, G.; Amin-Ahmadi, B.; Guo, Q.; Fan, G.; Guo, C.; Li, Z.; Zhang, D. Aligning graphene in bulk copper: Nacre-inspired nanolaminated architecture coupled with in-situ processing for enhanced mechanical properties and high electrical conductivity. *Carbon* **2017**, *117*, 65–74. [CrossRef]
44. Fan, G.; Jiang, Y.; Tan, Z.; Guo, Q.; Xiong, D.-b.; Su, Y.; Lin, R.; Hu, L.; Li, Z.; Zhang, D. Enhanced interfacial bonding and mechanical properties in CNT/Al composites fabricated by flake powder metallurgy. *Carbon* **2018**, *130*, 333–339. [CrossRef]
45. Zhong, Z.; Jiang, X.; Sun, H.; Pang, Y.; Wu, Z.; Yang, L. Effect of Cu alloying on the damping and compression properties of graphene nanoplatelets reinforced Al-30Zn-xCu alloy matrix composites. *Mater. Sci. Eng. A* **2024**, *915*, 147233. [CrossRef]
46. Bokobza, L.; Couzi, M.; Bruneel, J.-L. Raman spectroscopy of polymer–carbon nanomaterial composites. *Rubber Chem. Technol.* **2017**, *90*, 37–59. [CrossRef]
47. Yin, H.; Zhou, Y.; Liu, Q.; Wang, Y.; Tang, Z.; Yan, L. Corrosion behavior and mechanism of GH3535 alloy in NaOH, HF and H2O2 solution. *J. Alloys Compd.* **2021**, *887*, 161387. [CrossRef]
48. Patil, A.; Nartu, M.S.K.K.Y.; Ozdemir, F.; Banerjee, R.; Gupta, R.K.; Borkar, T. Enhancement of the mechanical properties of graphene nanoplatelet (GNP) reinforced nickel matrix nanocomposites. *Mater. Sci. Eng. A* **2021**, *817*, 141324. [CrossRef]

materials

MDPI

Article

Analytical Model of Temperature-Induced Deformation for Tunable Thermal Expansion Metamaterial

Ling Xiao [1,*], Yaxin Yao [1], Shuai Chen [2,*], Mengting Lai [1] and Guanghong Zhu [1]

[1] Department of Mechanics, Xi'an University of Science and Technology, Xi'an 710054, China; yaoyaxin219@163.com (Y.Y.); 18220528990@163.com (M.L.); zhuguanghong@xust.edu.cn (G.Z.)
[2] National Key Laboratory of Science and Technology on Advanced Composites in Special Environments, Harbin Institute of Technology, Harbin 150001, China
* Correspondence: xiaoling@xust.edu.cn (L.X.); chenshuai9312@hit.edu.cn (S.C.)

Abstract: Tunable thermal expansion metamaterials exhibit superior shock absorption performance in the field of high-precision equipment, but the applications are currently restricted by the unclear quantitative relationship of temperature-induced deformation. Herein, this work leverages the virtual work principle and the deformation geometric relationship to establish a generic temperature-induced deformation control model for bi-materials by utilizing the key variable coverage ratio under the condition of no deformation in the vertical direction. The feasible region regarding flexibility for the internal serpentine unit and lattice structure with different coverage ratios is given. The combination of the finite element and experimental methods is adopted to examine temperature-induced deformation, which presents tunable thermal expansion performances associated with the coverage ratio and temperature. This work, based on the established deformation coordination relationship of dual-material temperature-sensitive metamaterials, achieves temperature-induced deformation control and provides a reference for structural design adaptable in various working conditions such as vibration isolation and vibration reduction in complex engineering such as aerospace and so on. By strategically designing the coverage of the two structures within the specified range to maintain equivalent flexibility, the ultimate deformation of the serpentine unit is reduced by one-half due to deformation induced by temperature variations.

Keywords: tunable thermal expansion metamaterial (TTEM); virtual work principle; temperature-induced deformation; coverage ratio

1. Introduction

The increasing demand for thermal stability of systems under extreme environments, has led to metamaterials with tunable thermal expansion, negative stiffness [1,2], and extraordinary physical properties mainly through structural design becoming a research hotspot in the field of new materials [3]. Dual requirements for thermal and mechanical properties have motivated the design of stretch-dominated and bending-dominated [3–5] bi-material lattices induced by thermal exposure as an effective compensation for volume change.

Currently, many structures such as the slender straight or curved composite beam [6–8], the honeycomb structures including the simple honeycomb structure [9], the novel hybrid honeycomb structure by merging two hexagonal lattices [10], the lightweight 3D honeycomb structure by integrating concave double arrows with dual-material triangles [11], the 3D star-shaped honeycomb structure [12], and the 3D dual-material triangular cavity honeycomb structure [13] have been designed. However, the coefficient of thermal expansion (CTE) and negative stiffness are affected by geometric parameters, positional parameters, and material type. It is worth noting that a tunable metamaterial joint inspired by the molecular structure of carbon nanotubes has been proposed [14]. This joint can effectively absorb impact energy, reducing plastic deformation of material components. This further demonstrates the crucial influence of microstructural design on macroscopic mechanical performance and offers innovative perspectives and reference points for the design of the bi-material structure in the present research. Particularly, stress concentration on the triangular or star-shaped honeycomb structure is more likely to occur and has a larger numerical value.

Subsequently, the double-layer design based on lattice and mesh structures has gradually attracted widespread attention. Horseshoe metamaterials can achieve a notable negative Poisson's ratio and adjustable "J-shaped" stress–strain curves [15]. Significant progress has been made in the structural design of microstructures with geometric forms such as horseshoe [16–18], serpentine [19], and triangular [20] units. Subsequently, soft network materials with rotatable structural lattice nodes are designed by adding ring or disk-shaped structures to the horseshoe microstructure, which provide high stretchability while maintaining high strength [21]. Serpentine metamaterials exhibit high stretchability and temperature adaptability, and the strain and stiffness can be regulated by adjusting the width of the serpentine belt, arc radius, arc length, and arm length [19]. Metamaterials made up of triangular micro-units possess high stiffness and strength [22]. They are composed of trapezoidal micro-units that can achieve a broader range of thermal expansion tunability from positive to negative values in two orthogonal directions [20]. Four-dimensional printed intelligent lattice metamaterials can regulate the stiffness, energy absorption, and vibration damping effects by their altering structural parameters. However, their performance is significantly affected by temperature, where excessive temperatures lead to a decline in material strength, while excessively low temperatures impact energy absorption capability [23]. Furthermore, bifunctional metamaterials with different combinations of Poisson's ratio (PR) and coefficient of thermal expansion (CTE) can selectively deform under mechanical loading and cause dimensional changes under thermal loading, but the specific size cannot be estimated [24].

Furthermore, to meet the performance changes in materials in thermal environments and the specific requirements for thermal expansion behavior of engineering structures such as large space telescopes or antennas, it is necessary to fully consider the impact of temperature on the mechanical properties of materials from a theoretical perspective, which traditional materials cannot achieve. Recently, a thermally sensitive metamaterial assembled from highly sensitive thermostatic metal strips was produced, which exhibits significant shape transformation capabilities during the process of thermal energy conversion. The metamaterial is capable of achieving a design strain of 70% to 80% in just 5 s of heating, and reaches a thermal strain of approximately 30%, which is significantly higher than that of other bi-metallic metamaterials [25]. Based on the Euler beam theory, the temperature-dependent models for the thermal expansion and stiffness of metamaterials that consider the stability control and natural frequency enhancement factors have been established. It can accurately predict expansion-induced deformation and then improve structure reliability [20]. Furthermore, studies have investigated the photothermally activated energy transfer effects in single-walled carbon nanotube thin films using a vectorial two-wave mixing technique. These investigations have revealed that quantum and nonlinear optical phenomena, with varying degrees of involvement, lead to significant changes in the equivalent incident energy produced by the optical interaction of a single beam and two-wave mixing in terms of thermal conduction. This research underscores the importance of microstructural design on the macroscopic mechanical properties of materials and offers new insights and directions for the design of bi-material structures in this study [26]. Three-dimensional printed programmable horseshoe lattice structures based on a phase-evolution model may exhibit incomplete shape recovery due to non-uniform heating, which limits its application in areas requiring precise shape memory and uniform deformation recovery [27]. Two-dimensional expansion coefficients and stiffness models based on energy and stiffness matrix methods can be used to design metamaterials that can adapt to temperature changes and possess programmable CTE and stiffness [21]. Nevertheless, the deformation coordination relationship of metamaterials designed with double-layer assembly strategies in practical applications still needs further clarification. Additionally, there is insufficient guidance on how to precisely adjust these parameters in actual engineering to achieve non-deformation in specific directions.

According to the above information, the related research on this material mainly focuses on the simulation and experimentation of its thermal expansion behavior by changing the material's parameters. There is no complete design theory to calculate the temperature-induced deformation of bi-, tri-, or more metamaterial. Therefore, it is necessary to explore the intrinsic deformation relationships for a better optimization design.

Given the above requirements, this paper uses the bi-material serpentine unit as an example to conduct an in-depth analysis of its intrinsic deformation relationships. The displacement functions for the bi-material serpentine unit are established based on the principle of virtual work under different loads, respectively. The deformation coordination relationship between the internal serpentine unit and external lattice structure is obtained under the conditions of no deformation in the vertical direction. The feasible region with different coverage ratios is given and verified utilizing the finite element method and experimentation, which can be used for the structural design of temperature-sensitive bi-metamaterials.

2. Models and Methods

2.1. Geometry of the TTEM

The bi-material serpentine unit in TTEM is shown in Figure 1a,b, where it evolved from the horseshoe-shaped microstructure (yellow part). The elementary unit for the TTEM, as shown in Figure 1d, is an internal serpentine unit which is superimposed by an internal red actuation layer and an external yellow frame layer. There is a difference in the thermal expansion between the inner red actuation layer and the outer yellow frame layer. In this structure, the thermal expansion coefficient of the inner red actuation layer is higher than that of the outer yellow frame layer. Under the conditions of maintaining the longitudinal displacement unchanged, the material is capable of free deformation in the cross-sectional plane under the influence of temperature. Given that the length in the Z-direction is relatively longer compared to the X-direction, the deformation in the Z-direction is relatively small compared to the initial length. Based on this analysis, this study neglects the deformation effects in the Z-direction. When a temperature change ΔT occurs, due to the difference in the coefficients of thermal expansion between the two material layers, the semicircular double-layer beam undergoes local bending or rotation. According to the Euler–Bernoulli beam theory, the relationship between the linear strain and temperature has been analyzed [28].

$$\varepsilon_T = \alpha_{meta}\Delta T = \beta(E, E_2, \frac{t_1}{R_2}, \frac{t_2}{R_2})f(\theta_1, r)(\alpha_2 - \alpha_1)\Delta T \quad (1)$$

where

$$\beta(E, E_2, \frac{t_1}{R_2}, \frac{t_2}{R_2}) = \frac{6(\frac{t_1}{R_2} + \frac{t_2}{R_2})}{\frac{Et_1}{E_2t_2}(\frac{t_1}{R_2})^2 + \frac{E_2t_2}{Et_1}(\frac{t_2}{R_2})^2 + 4(\frac{t_1}{R_2})^2 + 4(\frac{t_2}{R_2})^2 + 6\frac{t_1}{R_2}\frac{t_2}{R_2}} \quad (2)$$

$$f(\theta_1, r) = \frac{2\sin(r\theta_1/2) - r\theta_1\cos(\theta_1/2)}{2\sin(\theta_1/2)} \quad (3)$$

where t_1, E, α_1, θ_1, and R_2 represent the wall thickness, elastic modulus, coefficient of thermal expansion, angle, and inner diameter of the frame layer; t_2, E_2, and α_2 represent the wall thickness, elastic modulus, and thermal expansion coefficient of the excitation layer; $r(0 \leq r \leq 1)$ is the ratio of the incentive layer material to the frame layer. As shown in Equations (1)–(3), the effective thermal expansion coefficient of metamaterials α_{meta} is highly dependent on the geometric parameters t_1/R_2, t_2/R_2, r, and θ_1. Thus, the suitable equivalent thermal expansion coefficient can be obtained by adjusting the geometric parameters, and then the temperature-induced deformation can be restricted according to actual needs.

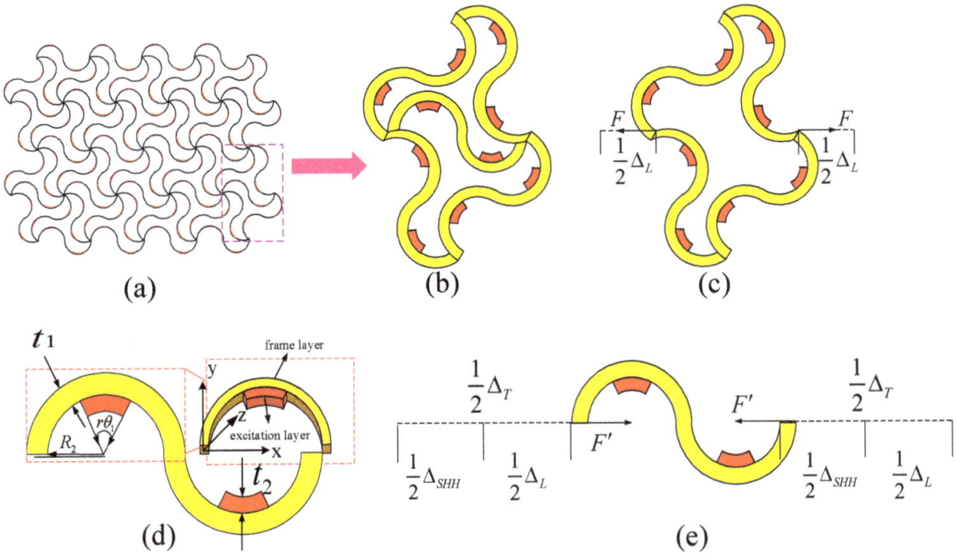

Figure 1. Structures and parameters of the tunable thermal expansion metamaterial. (**a**) Schematic illustration of the tunable thermal expansion metamaterial; (**b**) a 2D unit cell; (**c**) thermal expansion deformation of the lattice structure; (**d**) bi-material serpentine unit; and (**e**) thermal expansion deformation of the serpentine unit.

2.2. *Displacement Functions for the Bi-Material Serpentine Unit*

The tunable thermal expansion metamaterials deform with temperature changes and are subjected to complex forces. In this section, a bi-material serpentine unit is considered for analysis which is the smallest structure of the given material. According to the principle of simplification of the force system, the complex force system can be simplified into three basic loads including transverse force, longitudinal force, and bending moment. In Figure 2, the bi-material serpentine unit is divided into I–IV regions. θ_0 and θ_t are the start and end angles of the actuation layer, respectively. Among them, the solid line represents the load, and the dashed line represents the displacement. When the serpentine unit is subjected to a transverse force F_H, it produces transverse deformation Δ_{HH} and longitudinal deformation

Δ_{ZH}, as shown in Figure 2a. When the serpentine unit is subjected to a longitudinal force F_Z or bending moment M, deformations are generated, as shown in Figure 2b,c.

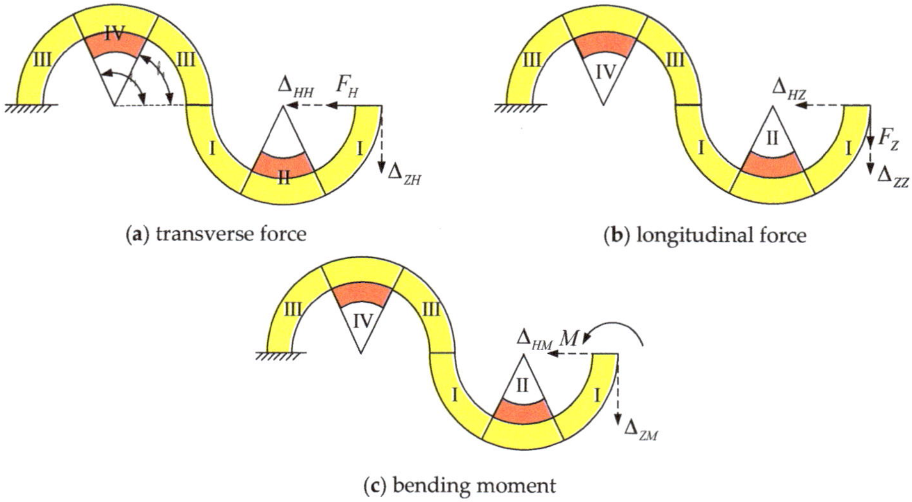

Figure 2. Mechanical models for the serpentine unit.

Based on the principles of virtual work, the displacement of the free end under different loads is calculated by Equation (4).

$$
\begin{aligned}
\Delta &= \Delta_I + \Delta_{II} + \Delta_{III} + \Delta_{IV} \\
&= \left(\Sigma\int \frac{\overline{M_1}M_1}{ER_CS}ds + \Sigma\int \frac{k\overline{F_{S1}}F_{S1}}{GA}ds + \Sigma\int \frac{\overline{F_{N1}}F_{N1}}{EA}ds + \Sigma\int \frac{\overline{F_{N1}}M_1}{EAR_C}ds\right. \\
&\left.+\Sigma\int \frac{\overline{M_1}F_{N1}}{EAR_C}ds\right) + \left(\Sigma\int \frac{\overline{M_1}M_1}{E_dR_{Cd}S_d}ds + \Sigma\int \frac{k\overline{F_{S1}}F_{S1}}{G_dA_d}ds + \Sigma\int \frac{\overline{F_{N1}}F_{N1}}{E_dA_d}ds\right. \\
&\left.+\Sigma\int \frac{\overline{F_{N1}}M_1}{E_dR_{Cd}}ds + \Sigma\int \frac{\overline{M_1}F_{N1}}{E_dA_dR_{Cd}}ds\right) + \left(\Sigma\int \frac{\overline{M_2}M_2}{ER_CS}ds + \Sigma\int \frac{k\overline{F_{S2}}F_{S2}}{GA}ds\right. \\
&\left.+\Sigma\int \frac{\overline{F_{N2}}F_{N2}}{EA}ds + \Sigma\int \frac{\overline{F_{N2}}M_2}{EAR_C}ds + \Sigma\int \frac{\overline{M_2}F_{N2}}{EAR_C}ds\right) + \left(\Sigma\int \frac{\overline{M_2}M_2}{E_dR_{Cd}S_d}ds\right. \\
&\left.+\Sigma\int \frac{k\overline{F_{S2}}F_{S2}}{G_dA_d}ds + \Sigma\int \frac{\overline{F_{N2}}F_{N2}}{E_dA_d}ds + \Sigma\int \frac{\overline{F_{N2}}M_2}{E_dR_{Cd}}ds + \Sigma\int \frac{\overline{M_2}F_{N2}}{E_dA_dR_{Cd}}ds\right)
\end{aligned}
\tag{4}
$$

where M_1, F_{N1} and F_{S1} are the bending moment, axial force, and shear force of any section on the right segment of the serpentine unit when the free end is subjected to an external load, while $\overline{M}_1$, $\overline{F}_{N1}$ and $\overline{F}_{S1}$ are the corresponding internal forces of the same section. M_2, F_{N2} and F_{S2} are the corresponding forces of any section on the left segment, while $\overline{M}_2$, $\overline{F}_{N2}$ and $\overline{F}_{S2}$ are the corresponding internal forces of the same section. G, E, G_2, and E_2 are the shear and elastic modulus of the frame layer material and the excitation layer material, respectively. $G_d = (AG + A_2G_2)/(A + A_2)$ and $E_d = (AE + A_2E_2)/(A + A_2)$ represent the partial shear strength and elastic modulus of the serpentine unit, k, A, $S = Ay = A[R_C - t_1/\ln(R_1/R_2)]$, R_c, and A_2 are the correction factors for the uneven distribution of shear stress along the section, the cross-sectional area of the frame layer, the area moment of the frame layer section to the central axis, the radius of the centroid of the frame layer section, and the cross-sectional area of the excitation layer, respectively. $A_d = A + A_2$ is the composite cross-sectional area. $S_d = (A + A_2)\{R_{Cd} - (t_1 + t_2)/[\ln(R_1/(R_2 - t_2))]\}$ is the area moment of the composite cross-section to the composite central axis. $R_{Cd} = R_C - t_2/2$ is the radius of the composite cross-section centroid.

It is known that both the transverse and the longitudinal displacements of the free end are produced when the serpentine unit is subjected to the transverse force shown in Figure 3. Under the action of the transverse force, the internal force equation of each part for the serpentine unit can be obtained by the section method.

$$\begin{array}{l} right\ section: \mathrm{M}_1 = F_H R_C \sin\theta, F_{S1} = -F_H \cos\theta, F_{N1} = -F_H \sin\theta \\ left\ section: \mathrm{M}_2 = -F_H R_C \sin\theta, F_{S2} = F_H \cos\theta, F_{N2} = -F_H \sin\theta \end{array} \quad (5)$$

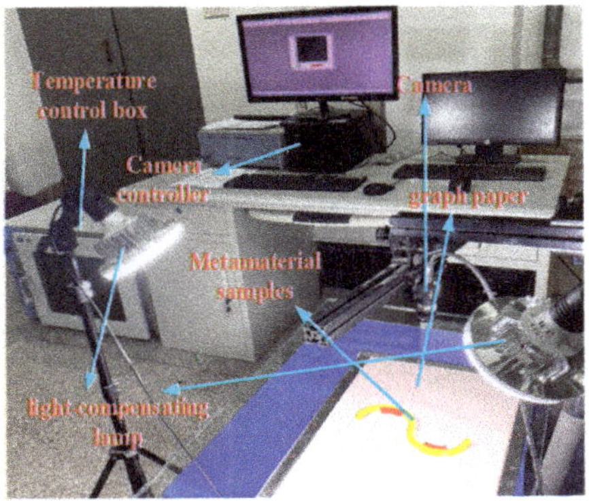

Figure 3. Heating and testing equipment.

The transverse and longitudinal displacements Δ_{HH}, Δ_{ZH} can be obtained by the section method under the action of virtual transverse and longitudinal forces. The internal forces of each part for the serpentine unit under the virtual transverse force is as follows:

$$\begin{array}{l} right\ section: \overline{\mathrm{M}}_1 = R_C \sin\theta, \overline{F}_{S1} = -\cos\theta, \overline{F}_{N1} = -\sin\theta \\ left\ section: \overline{\mathrm{M}}_2 = -R_C \sin\theta, \overline{F}_{S2} = \cos\theta, \overline{F}_{N2} = -\sin\theta \end{array} \quad (6)$$

Substituting Equations (5) and (6) into Equation (4), the transverse displacement Δ_{HH} is as follows:

$$\Delta_{HH} = K_{HH} F_H \quad (7)$$

The internal forces of each part for the serpentine unit under the virtual longitudinal force is as follows:

$$\begin{array}{l} right\ section: \overline{F}_{N1} = -\cos\theta, \overline{F}_{S1} = \sin\theta, \overline{\mathrm{M}}_1 = R_C(\cos\theta - 1) \\ left\ section: \overline{F}_{N2} = \cos\theta, \overline{F}_{S2} = \sin\theta, \overline{\mathrm{M}}_2 = R_C(\cos\theta - 3) \end{array} \quad (8)$$

Substituting Equations (5) and (8) into Equation (4), the longitudinal displacement Δ_{ZH} is as follows:

$$\Delta_{ZH} = K_{ZH} F_H \quad (9)$$

According to the above analysis, the transverse displacement flexibility K_{HH} and longitudinal displacement flexibility K_{ZH} of the serpentine unit under transverse force are as follows:

$$
\begin{aligned}
K_{HH} &= (\frac{R_C^2}{ES} - \frac{R_C}{EA})(\int_0^{\theta_0} \sin^2\theta d\theta + \int_{\theta_t}^{\pi} \sin^2\theta d\theta) + \frac{kR_C}{GA}(\int_0^{\theta_0} \cos^2\theta d\theta + \int_{\theta_t}^{\pi} \cos^2\theta d\theta) + \\
&(\frac{R_{Cd}^2}{E_d S_d} - \frac{R_{Cd}}{E_d A_d})\int_{\theta_0}^{\theta_t} \sin^2\theta d\theta + \frac{kR_{Cd}}{G_d A_d}\int_{\theta_0}^{\theta_t} \cos^2\theta d\theta + \\
&(\frac{R_C^2}{ES} + \frac{3R_C}{EA})(\int_0^{\theta_0} \sin^2\theta d\theta + \int_{\theta_t}^{\pi} \sin^2\theta d\theta) + \frac{kR_C}{GA}(\int_0^{\theta_0} \cos^2\theta d\theta + \int_{\theta_t}^{\pi} \cos^2\theta d\theta) + \\
&(\frac{R_{Cd}^2}{E_d S_d} + \frac{3R_{Cd}}{E_d A_d})\int_{\theta_0}^{\theta_t} \sin^2\theta d\theta + \frac{kR_{Cd}}{G_d A_d}\int_{\theta_0}^{\theta_t} \cos^2\theta d\theta
\end{aligned} \quad (10)
$$

$$
\begin{aligned}
K_{ZH} &= (\frac{R_C^2}{ES} - \frac{R_C}{EA})(\int_0^{\theta_0} \sin\theta(\cos\theta - 1)d\theta + \int_{\theta_t}^{\pi} \sin\theta(\cos\theta - 1)d\theta) + \\
&\frac{kR_C}{GA}(\int_0^{\theta_0} - \sin\theta\cos\theta d\theta + \int_{\theta_t}^{\pi} - \sin\theta\cos\theta d\theta) + \\
&(\frac{R_{Cd}^2}{E_d S_d} - \frac{R_{Cd}}{E_d A_d})\int_{\theta_0}^{\theta_t} \sin\theta(\cos\theta - 1)d\theta + \frac{kR_{Cd}}{G_d A_d}\int_{\theta_0}^{\theta_t} - \sin\theta\cos\theta d\theta + \\
&(\frac{R_C^2}{ES} + \frac{R_C}{EA})(\int_0^{\theta_0} - \sin\theta(\cos\theta - 3)d\theta + \int_{\theta_t}^{\pi} - \sin\theta(\cos\theta - 3)d\theta) + \\
&(\frac{kR_C}{GA} - \frac{2R_C}{EA})(\int_0^{\theta_0} \sin\theta\cos\theta d\theta + \int_{\theta_t}^{\pi} \sin\theta\cos\theta d\theta) + \\
&(\frac{R_{Cd}^2}{E_d S_d} + \frac{R_{Cd}}{E_d A_d})\int_{\theta_0}^{\theta_t} - \sin\theta(\cos\theta - 3)d\theta + (\frac{kR_{Cd}}{G_d A_d} - \frac{2R_{Cd}}{E_d A_d})\int_{\theta_0}^{\theta_t} \sin\theta\cos\theta d\theta
\end{aligned} \quad (11)
$$

Similarly, other parameters such as the transverse displacement flexibility K_{HZ}, longitudinal displacement flexibility K_{ZZ} under longitudinal force, as well as its transverse displacement flexibility K_{HM} and longitudinal displacement flexibility K_{ZM} under bending moment can be obtained. The details can be seen in Supplementary Materials.

2.3. Displacement Functions for the Lattice Structure

Four serpentine units are combined into a lattice structure, as shown in Figure 1b, where the upper and lower ends of the structure are constrained by fixed ends, and the left and right ends are subjected to a pair of equal and opposite forces. The total deformation at the two endpoints in Equation (12) is obtained by the superposition principle and the details can be seen in Supplementary Materials.

$$
\begin{aligned}
\Delta_L &= \Delta_{lx} - \Delta_{rx} \\
&= (K_{lSHH} + K_{rSHH})(\tfrac{\sqrt{3}}{4}F_1 + \tfrac{1}{8}F) + (-K_{lSHZ} + K_{rSHZ})(\tfrac{1}{4}F_1 - \tfrac{\sqrt{3}}{8}F) + \\
&(K_{lSZH} - K_{rSZH})(\tfrac{3}{4}F_1 + \tfrac{\sqrt{3}}{8}F) + (-K_{lSZZ} - K_{rSZZ})(\tfrac{\sqrt{3}}{4}F_1 - \tfrac{3}{8}F) + \\
&[\tfrac{1}{2}\left(K_{lSHM} - K_{rSHM}\right) + \tfrac{\sqrt{3}}{2}(K_{lSZM} + K_{rSZM})]M
\end{aligned} \quad (12)
$$

$F = 1$ and Δ_L represent the displacement of the lattice structure under a pair of unit loads with equal size and opposite directions, that is, the flexibility K_J for the lattice structure.

2.4. Deformation Coordination Relationship for the TTEM

In order to control the temperature-induced deformation of the TTEM, it is necessary to study the relationship between the deformation and coverage (the ratio of the curvature of the excitation layer to the frame layer) of the internal serpentine unit and the external lattice structure. To simplify the calculation, it is assumed that the temperature-induced deformation of the serpentine unit in the external lattice is canceled with each other. For tunable thermal expansion metamaterial in Figure 1c, when a force F induced by the thermal expansion in the serpentine unit acts on the lattice structure, it will deform Δ_L as shown in Figure 1c.

$$
\Delta_L = K_J F \quad (13)
$$

In the internal serpentine unit, firstly, it expands Δ_T when heated, then shortened Δ_{SHH} by the lattice structure reaction force F', and its total deformation Δ_L is shown in Figure 1e.

$$\Delta_{SHH} = K_{SHH}F' \tag{14}$$

where $\Delta_{SHH} = \Delta_T - \Delta_L$, $F = F'$, K_J and K_{SHH} represent the displacement flexibility of the lattice structure and the serpentine unit, respectively. Combining Equation (13) with Equation (14) produces the following equation:

$$\frac{\Delta_L}{K_J} = \frac{\Delta_T - \Delta_L}{K_{SHH}} \tag{15}$$

From Equation (15), it can be seen that the flexibility of lattice structure is equal to that of serpentine unit when the final deformation of serpentine unit is only half of the temperature-induced deformation, namely, $\Delta_L = \Delta_T/2$. Thus, the deformation coordination relationship as shown in Equation (16).

$$K_J = K_{SHH} \tag{16}$$

3. Results and Discussion

3.1. Material Preparation

In this study, we choose polymethyl methacrylate (PMMA, $E_{PMMA} = 2500$ Mpa, $\alpha_{PMMA} = 7.0 \times 10^{-5}\text{K}^{-1}$) as the internal red actuation layer, and polyimide (PI, $E_{PI} = 2500$ Mpa, $\alpha_{PI} = 3.0 \times 10^{-5}\text{K}^{-1}$) and polylactide (PLA, $E_{PLA} = 2500$ Mpa, $\alpha_{PLA} = 5.0 \times 10^{-5}\text{K}^{-1}$) as the external yellow frame layer. The above three materials are divided into two groups, in which the first group (PI+PMMA) is abbreviated as G.1 and the second group (PLA+PMMA) is abbreviated as G.2. All of these resin materials have the same mechanical property, such as Young's modulus $E = 2500$ MPa and Poisson's ratio $\mu = 0.3$.

The printing, heating, and testing equipment are illustrated in Figure 3. The left end of the serpentine unit as well as the upper and lower ends of the tunable thermal expansion metamaterial, are fixed. The heating rate is set to 1.5833 °C/min from 25 °C to 95 °C and a constant temperature is maintained for 1 h.

3.2. Deformation Coordination Relationship

Figure 4 clearly depicts the trend of lateral displacement (blue) and longitudinal displacement (red) as the internal radius R2 increases at a fixed coverage rate of 50%. It is evident from the figure that both lateral and longitudinal displacements exhibit a stable linear growth as R_2 increases. This phenomenon indicates that under the specific conditions of this study, increasing R_2 leads to an increase in the structural displacement. This linear relationship is crucial for understanding and predicting the behavior of structures under different design parameters.

Figure 5 meticulously delineates the functional relationship between the X-direction deformation (in meters) and the coverage rate (in percentage) based on the predictive outcomes of an analytical model. The overall trend of the curve indicates a decrease in the X-direction displacement as the coverage rate increases, providing a clear quantitative basis for understanding the impact of the coverage rate on the deformation behavior of materials or structures.

Figure 6 shows the design results of the deformation coordination relationship of the tunable thermal expansion metamaterials with a Young's modulus of 2500 Mpa and a Poisson's ratio of 0.3. Flexibility analysis for the bi-material serpentine unit and the lattice structure are shown in Supplementary Materials Sections 3 and 4. The feasible region can be determined by the flexibility curve to make the lattice structure's flexibility K_J equal to the flexibility K_{SHH} of the serpentine unit, as shown in the shadow range. When the required compliance of the lattice structure K_J and compliance of the serpentine unit K_{SHH} both fall within the feasible range of 0.75×10^{-2}mm/N to 1.5×10^{-2}mm/N, it is necessary to determine the corresponding range of coverage rates. By precisely controlling this range of coverage rates, the desired flexibility can be achieved for both structures, thereby effectively controlling the temperature-induced deformation.

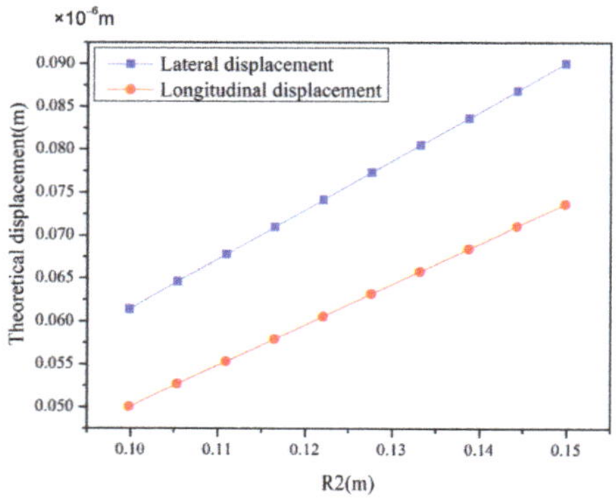

Figure 4. Effect of R_2 variation on lateral and longitudinal displacements at 50% coverage rate.

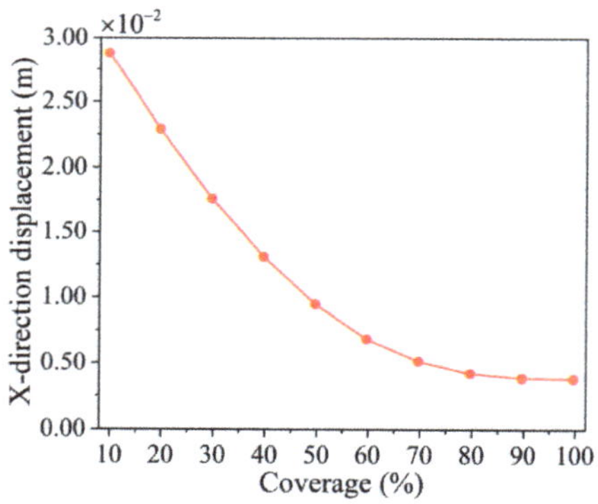

Figure 5. Relationship between lattice structure's x-direction displacement and coverage rate.

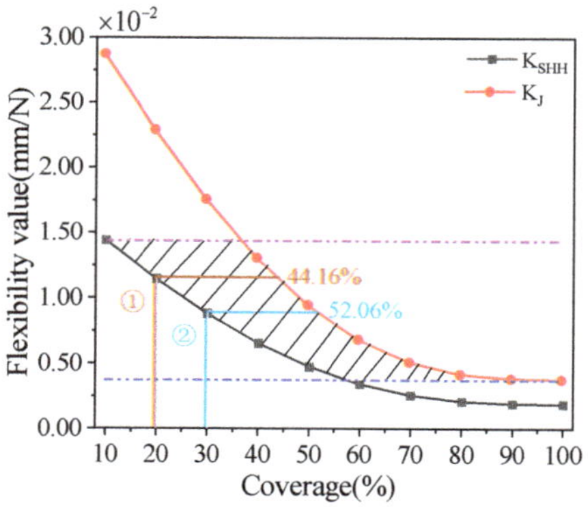

Figure 6. The relationship between flexibility and coverage of the internal serpentine unit and external lattice structure.

3.3. Simulation Verification

In order to verify the deformation coordination relationship of the tunable thermal expansion metamaterial, the 2D finite element method is used to analyze two cases, and the specific coverage is shown in Figure 6 ① and ②. The contact surface between the excitation layer and the frame layer adopts a binding method. And the free mesh division with a mesh size of 1 mm is used. The fixed-end constraints are applied to the upper and lower ends of the lattice structure, and the temperature change in the internal serpentine is set between 25 °C and 95 °C. The first group (PI+PMMA) is selected to analyze. For case ①, when the temperature reaches 95 °C, the temperature deformation of the serpentine unit and the final deformation of the left and right ends are 4.74 mm (Figure 7a) and 2.33 mm (Figure 7b), respectively. In the same way, for case ②, the temperature deformation and the final deformation of the serpentine unit are 6.51 mm (Figure 7c) and 3.30 mm (Figure 7d), respectively. Compared with the analytic model, the error of the total deformation is less than 2%.

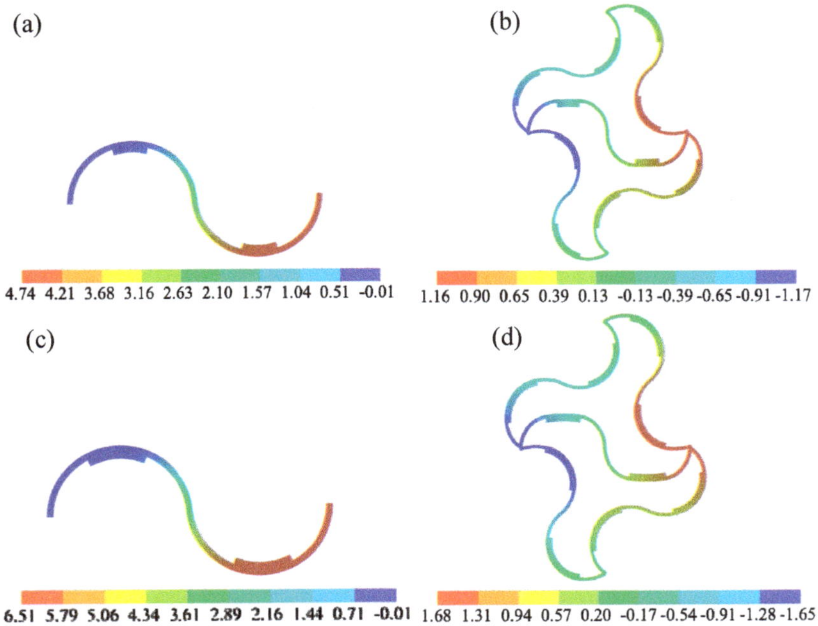

Figure 7. Temperature deformation for the serpentine unit under the coverages of 20% (**a**) and 30% (**b**). The final deformation for the serpentine unit under the coverages of 20% (**c**,**d**).

The remaining sets of validation data are shown in Table 1, where, regardless of G.1 or G.2, the temperature deformation is nearly twice that of the final deformation for the serpentine unit in the metamaterial. In addition, because the interaction between the excitation layer and the framework is ignored after selecting coverage from the shadow range outward, the error gradually increases. This mainly stems from the superposition of errors in the analysis model of the serpentine unit and lattice structure. From Table 1, we can choose a specific coverage rate to achieve better control results. All the simulation results verify that the design method proposed in this paper is basically feasible and can be used before the experiment.

Table 1. Deformation of two groups of the serpentine units in G.1 and G.2.

Coverage of the Serpentine Unit (%)	Lattice Structure Coverage (%)	Temperature Deformation of the Serpentine Unit (mm)		Simulation Value of the Final Deformation of the Serpentine Unit (mm)		Analytic Model of the Final Deformation of the Serpentine Unit (mm)	
		G.1	G.2	G.1	G.2	G.1	G.2
10	36.97	2.93	2.88	1.26	1.16	1.47	1.44
20	44.16	4.74	3.78	2.33	1.81	2.37	1.89
30	52.06	6.51	4.67	3.30	2.29	3.26	2.34
40	61.62	8.18	5.51	4.32	2.85	4.09	2.75
50	74.16	9.62	6.53	5.19	3.35	4.81	3.11
57.24	100	10.60	6.71	4.64	2.92	5.30	3.35

Figure 8 provides a comparative analysis of the error percentages for two materials, G.1 (PI+PMMA) and G.2 (PLA+PMMA), across different coverage rates of the serpentine unit. For both materials, the error percentages initially decrease and then increase as the coverage rate of the serpentine unit rises. At a coverage rate of 10%, the error percentage for material G.2 peaks at approximately 19%, while for material G.1, it is around 16% under the same conditions. At a coverage rate of 60%, material G.1 achieves an error percentage of about 14%, whereas material G.2 records an error percentage of approximately 13%. In summary, both materials demonstrate good performance under the same coverage conditions.

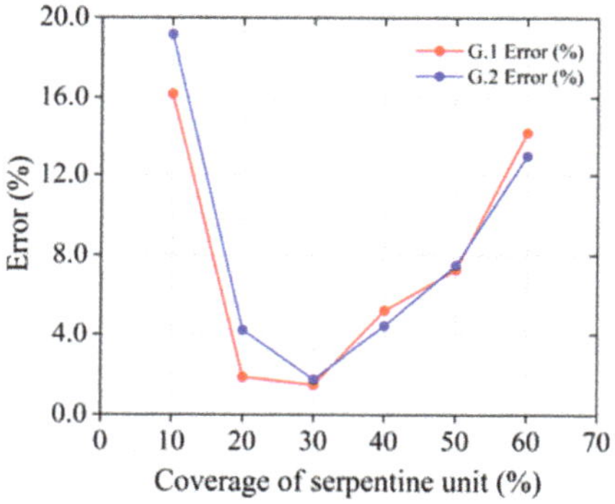

Figure 8. Error comparison between G.1 (PI+PMMA) and G.2 (PLA+PMMA).

3.4. Experimental Verification

To avoid measurement errors caused by temperature loss during the measurement process, multiple experiments were conducted on each case and the average measurement value was calculated. Figure 9a,c show the serpentine unit specimens with 20% and 30% coverage at room temperature, respectively. When the temperature changed from 25 °C to 95 °C during the heating process, the temperature deformation was 3.7 mm and 4.3 mm, respectively, as shown in Figure 9b,d. For samples a-2 and b-2 in Figure 10, the total temperature-induced deformation at the left and right ends is 2.1 mm and 2.5 mm in the same temperature range, respectively, as shown in Figure 10a-3,b-3. The reason for the error may come from the loss of temperature and the model difference between the analytical model (2D) and the experiment (3D).

The experimental results indicate that the temperature-induced deformation of the serpentine unit is nearly twice that of the final deformation of the metamaterial, which confirms the temperature-induced deformation coordination relationship proposed in this paper. The finite element method can be employed to calculate the specific deformation dimensions within the feasible region. Analytical and simulation methods can be applied to the structural design of temperature-sensitive bi-metamaterials, and can even be used to precisely adjust parameters in practical engineering to achieve non-deformation in specific directions.

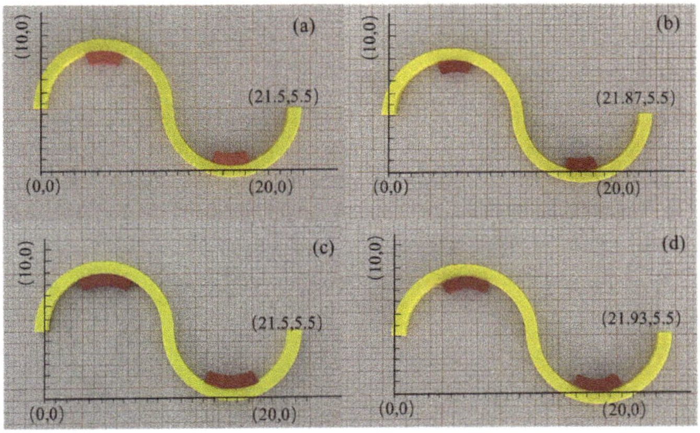

Figure 9. Thermal expansion deformation of the serpentines unit. (**a**) Room temperature (20% coverage); (**b**) after thermal deformation (20% coverage); (**c**) room temperature (30% coverage); (**d**) sample after thermal deformation (30% coverage).

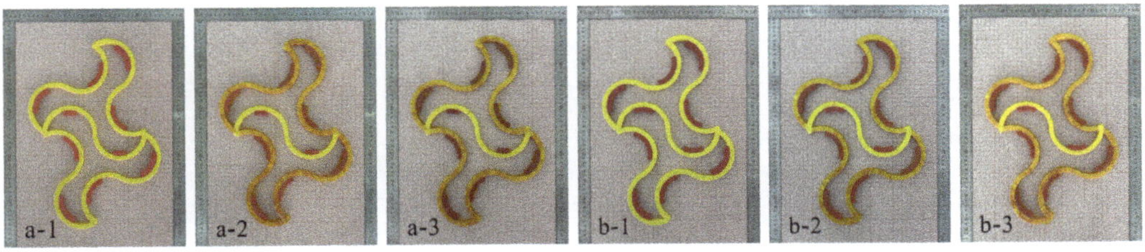

Figure 10. Tunable thermal expansion metamaterial deformation (**a-1**) at room temperature (Case 1); (**a-2**) room temperature+insulation treatment (Case 1); (**a-3**) after thermal deformation (Case 1); (**b-1**) room temperature (Case 2); (**b-2**) room temperature+insulation treatment (Case 2); (**b-3**) after thermal deformation (Case 2).

4. Conclusions

In this paper, the deformation characteristics of tunable thermal expansion metamaterials are explored through analytical and simulation approaches, leading to the following conclusions:

1. Displacement flexibility functions for the serpentine units and lattice structure within tunable thermal expansion metamaterials are formulated using the principle of virtual work and the method of superposition.
2. It is discovered that the internal serpentine unit and the external lattice structure exhibit equivalent flexibility when the final deformation of the serpentine unit is half of its temperature-induced deformation. Based on this finding, this study presents a method for analyzing the deformation coordination relationship between the lattice and serpentine unit coverage rates in metamaterials.
3. The proposed deformation coordination relationship is validated through finite element analysis and experimental methods. With an internal unit coverage rate of 30% and an external lattice structure coverage rate of 52.06%, the prediction error is maintained within 1.47%. These research outcomes provide a theoretical foundation and

design guidance for precise temperature-induced deformation control of bi-material serpentine network structural units.

Supplementary Materials: The following supporting information can be downloaded at: https://www.mdpi.com/article/10.3390/ma18030532/s1, Figure S1. Forces on the lattice structure. Figure S2. Relationship between coverage and flexibility value for the serpentine unit. Figure S3. Relationship between coverage and flexibility value for lattice structure of K_J.

Author Contributions: Conceptualization, L.X. and M.L.; Methodology, Y.Y. and M.L.; Validation, Y.Y.; Formal analysis, L.X.; Data curation, S.C.; Writing—original draft, L.X. and Y.Y.; Writing—review & editing, L.X., Y.Y. and S.C.; Visualization, Y.Y.; Supervision, L.X., S.C. and G.Z.; Funding acquisition, L.X. All authors have read and agreed to the published version of the manuscript.

Funding: The authors disclosed receipt of the following financial support for the research, authorship, and/or publication of this article. This study was supported by the National Natural Science Foundation of China (Grant No. 52275271, 12002267).

Institutional Review Board Statement: Not applicable.

Informed Consent Statement: Not applicable.

Data Availability Statement: The original contributions presented in this study are included in the article/supplementary material. Further inquiries can be directed to the corresponding author.

Conflicts of Interest: The authors declare no conflicts of interest.

References

1. Liu, T.; Deng, R.; Zhang, Y.; Yang, J.; Cai, J. Composite negative stiffness structure with tunable and temperature-dependent properties induced by viscoelastic and shape-memory materials. *Compos. Commun.* **2024**, *48*, 101937. [CrossRef]
2. Liu, S.; Li, F.; Peng, J.; Zhang, X. Universal model describing the negative thermal expansion coefficients of bending-type two-dimensional metamaterials with chiral/anti-chiral structures. *Compos. Commun.* **2023**, *39*, 101559. [CrossRef]
3. Wei, K.; Peng, Y.; Wang, K.; Duan, S.; Yang, X.; Wen, W. Three dimensional lightweight lattice structures with large positive, zero and negative thermal expansion. *Compos. Struct.* **2018**, *188*, 287–296. [CrossRef]
4. He, X.; Yu, J.; Xie, Y. Bi-material re-entrant triangle cellular structures incorporating tailorable thermal expansion and tunable Poisson's ratio. *J. Mech. Robot.* **2019**, *11*, 061003. [CrossRef]
5. Wei, K.; Peng, Y.; Qu, Z.; Duan, S.; Yang, X.; Wen, W. Lightweight composite lattice cylindrical shells with novel character of tailorable thermal expansion. *Int. J. Mech. Sci.* **2018**, *137*, 77–85. [CrossRef]
6. Timoshenko, S. Analysis of bi-metal thermostats. *Josa* **1925**, *11*, 233–255. [CrossRef]
7. Lakes, R. Cellular solid structures with unbounded thermal expansion. *J. Mater. Sci. Lett.* **1996**, *15*, 475–477. [CrossRef]
8. Lakes, R. Cellular solids with tunable positive or negative thermal expansion of unbounded magnitude. *Appl. Phys. Lett.* **2007**, *90*, 359. [CrossRef]
9. Zheng, B.B.; Fu, M.H.; Li, W.H.; Hu, L.L. A novel re-entrant honeycomb of negative thermal expansion. *Smart Mater. Struct.* **2018**, *27*, 085005. [CrossRef]
10. Peng, X.L.; Bargmann, S. A novel hybrid-honeycomb structure: Enhanced stiffness, tunable auxeticity and negative thermal expansion. *Int. J. Mech. Sci.* **2021**, *190*, 106021. [CrossRef]
11. Peng, Y.; Wei, K.; Mei, M.; Yang, X.; Fang, D. Simultaneously program thermal expansion and Poisson's ratio in three dimensional mechanical metamaterial. *Compos. Struct.* **2021**, *262*, 113365. [CrossRef]
12. Liu, H.T.; Xu, N.; Wang, Y.B.; Wang, L. Three-dimensional enhanced star-shaped honeycombs with negative thermal expansion. *Compos. Struct.* **2022**, *279*, 114772. [CrossRef]
13. Liu, K.J.; Liu, H.T.; Zhen, D. Mechanical and bandgap properties of 3D bi-material triangle re-entrant honeycomb. *Int. J. Mech. Sci.* **2024**, *261*, 108664. [CrossRef]
14. Giannopoulos, G.I.; Georgantzinos, S.K. A tunable metamaterial joint for mechanical shock applications inspired by carbon nanotubes. *Appl. Sci.* **2021**, *11*, 11139. [CrossRef]
15. Liu, J.; Zhu, X.; Shen, Z.; Zhang, Y. Imperfection sensitivity of mechanical properties in soft network materials with horseshoe microstructures. *Acta Mech. Sin.* **2021**, *37*, 1050–1062. [CrossRef]
16. Liu, G.; Zheng, Z.; Zhao, R.; Yue, X. Horseshoe Lattice Property-Structure Inverse Design Based on Deep Learning. *Mater. Trans.* **2024**, *65*, 308–317. [CrossRef]

17. Zhou, J.; Chang, J.; Song, X.; Zhang, Y. Bio-inspired design and unusual mechanical properties of 3D horseshoe-shaped soft network metamaterials. *Compos. Part B Eng.* **2024**, *275*, 111284. [CrossRef]
18. Widlund, T.; Yang, S.; Hsu, Y.Y.; Lu, N. Stretchability and compliance of freestanding serpentine-shaped ribbons. *Int. J. Solids Struct.* **2014**, *51*, 4026–4037. [CrossRef]
19. Chen, J.; Wang, H.; Wang, K.; Wei, Z.; Xu, W.; Wei, K. Mechanical performances and coupling design for the mechanical metamaterials with tailorable thermal expansion. *Mech. Mater.* **2022**, *165*, 104176. [CrossRef]
20. Zhang, Q.; Sun, Y. Novel metamaterial structures with negative thermal expansion and tunable mechanical properties. *Int. J. Mech. Sci.* **2024**, *261*, 108692. [CrossRef]
21. Li, J.; Liu, H.T.; Zhang, Z.Y. Stiffness characteristics for bi-directional tunable thermal expansion metamaterial based on bi-material triangular unit. *Int. J. Mech. Sci.* **2023**, *241*, 107983. [CrossRef]
22. Liu, J.; Yan, D.; Zhang, Y. Mechanics of unusual soft network materials with rotatable structural nodes. *J. Mech. Phys. Solids* **2021**, *146*, 104210. [CrossRef]
23. Zhang, X.; Han, Y.; Zhu, M.; Chu, Y.; Li, W.; Zhang, Y.; Zhang, Y.; Luo, J.; Tao, R.; Qi, J. Bio-inspired 4D printed intelligent lattice metamaterials with tunable mechanical property. *Int. J. Mech. Sci.* **2024**, *272*, 109198. [CrossRef]
24. Han, Z.; Xiao, X.; Chen, J.; Wei, K.; Wang, Z.; Yang, X.; Fang, D. Bifunctional metamaterials incorporating unusual geminations of Poisson's ratio and coefficient of thermal expansion. *ACS Appl. Mater. Interfaces* **2022**, *14*, 50068–50078. [CrossRef] [PubMed]
25. Zhang, Y.; Jiang, W.Z.; Pan, Y.; Teng, X.C.; Xu, H.H.; Yan, H.; Ni, X.H.; Dong, J.; Han, D.; Chen, W.Q.; et al. Temperature-responsive metamaterials made of highly sensitive thermostat metal strips. *Sci. Adv.* **2024**, *10*, eads0892. [CrossRef]
26. Hernández-Acosta, M.A.; Martines-Arano, H.; Soto-Ruvalcaba, L.; Martínez-González, C.L.; Martínez-Gutiérrez, H.; Torres-Torres, C. Fractional thermal transport and twisted light induced by an optical two-wave mixing in single-wall carbon nanotubes. *Int. J. Therm. Sci.* **2020**, *147*, 106136. [CrossRef]
27. Wang, D.; Xu, H.; Wang, J.; Jiang, C.; Zhu, X.; Ge, Q.; Gu, G. Design of 3D printed programmable horseshoe lattice structures based on a phase-evolution model. *ACS Appl. Mater. Interfaces* **2020**, *12*, 22146–22156. [CrossRef] [PubMed]
28. Ni, X.; Guo, X.; Li, J.; Huang, Y.; Zhang, Y.; Rogers, J.A. 2D mechanical metamaterials with widely tunable unusual modes of thermal expansion. *Adv. Mater.* **2019**, *31*, 1905405. [CrossRef]

materials

MDPI

Article

Molecular Dynamics Insights into Mechanical Stability, Elastic Properties, and Fracture Behavior of PHOTH-Graphene

Qing Peng [1,2,3,*], Gen Chen [2,4,5], Zeyu Huang [2,4,5], Xue Chen [2], Ao Li [2], Xintian Cai [2,6], Yuqiang Zhang [4,5], Xiao-Jia Chen [1,*] and Zhongwei Hu [4,5,*]

1 School of Science, Harbin Institute of Technology, Shenzhen 518055, China
2 State Key Laboratory of Nonlinear Mechanics, Institute of Mechanics, Chinese Academy of Sciences, Beijing 100190, China; 22013080001@stu.hqu.edu.cn (G.C.); 22014080032@stu.hqu.edu.cn (Z.H.); chenxue@imech.ac.cn (X.C.); ao@youzi.com (A.L.); caixintian@whu.edu.cn (X.C.)
3 Guangdong Aerospace Research Academy, Guangzhou 511458, China
4 Institute of Manufacturing Engineering, Huaqiao University, Xiamen 361021, China; 19013080047@stu.hqu.edu.cn
5 Institute of Mechanical Engineering and Automation, Huaqiao University, Xiamen 361021, China
6 Hubei Key Laboratory of Electronic Manufacturing and Packaging Integration, Wuhan University, Wuhan 430072, China
* Correspondence: pengqing@imech.ac.cn (Q.P.); xjchen@hit.edu.cn (X.-J.C.); huzhongwei@hqu.edu.cn (Z.H.)

Abstract: PHOTH-graphene is a newly predicted 2D carbon material with a low-energy structure. However, its mechanical stability and fracture properties are still elusive. The mechanical stability, elastic, and fracture properties of PHOTH-graphene were investigated using classical molecular dynamics (MD) simulations equipped with REBO potential in this study. The influence of orientation and temperature on mechanical properties was evaluated. Specifically, the Young's modulus, toughness, and ultimate stress and strain varied by −26.14%, 36.46%, 29.04%, and 25.12%, respectively, when comparing the armchair direction to the zigzag direction. The percentage reduction in ultimate stress, ultimate strain, and toughness of the material in both directions after a temperature increase of 1000 K (from 200 K to 1200 K) ranged from 56.69% to 91.80%, and the Young's modulus was reduced by 13.63% and 7.25% in both directions, respectively, with Young's modulus showing lower sensitivity. Defects usually weaken the material's strength, but adding random point defects in the range of 3% to 5% significantly increases the ultimate strain of the material. Furthermore, hydrogen atom adsorption induces crack expansion to occur earlier, and the crack tip without hydrogen atom adsorption just began to expand when the strain was 0.135, while the crack tip with hydrogen atom adsorption had already undergone significant expansion. This study provides a reference for the possible future practical application of PHOTH-graphene in terms of mechanical properties and fracture failure.

Keywords: PHOTH-graphene; mechanical property; MD; defects; hydrogen functionalization

1. Introduction

Two-dimensional graphene exhibits excellent properties and is employed in many applications, including electronic components, hydrogen storage, batteries, and desalination membranes [1–7]. The Young's modulus of monolayer graphene is up to 1 TPa, the ultimate stress can reach 130 GPa, and the ultimate strain can reach more than 0.20 [8–12]. The extraordinary mechanical properties of graphene have attracted researchers to conduct a series of studies on its mechanical properties and fracture failure mechanisms. These studies have included an investigation of the material's anisotropy and an examination of the influences of temperature, defects, and functional group adsorption on both mechanical properties and fracture behavior. The strength of graphene is compromised at elevated temperatures [13,14], with the material exhibiting brittle fracture behavior [15]. Even at ultrahigh temperatures of 4000 K~5000 K, the fracture mode of graphene remains brittle [16].

Defects are inevitable in the preparation and engineering applications of graphene, and common forms of defects include Stone–Thrower–Wales (STW) defects, vacancy defects, crack and line defects (dislocations and grain boundaries) [17–19], etc. The introduction of defects usually weakens the strength of the material [20–22], but the mechanical properties of graphene can also be improved through reasonable design of the defects to control anisotropy and enhance toughness [17,23–26]. Carbon materials, represented by graphene, are considered to be excellent solid-state hydrogen storage materials due to their high porosity [27,28]. Hydrogen atoms in the external environment change the electronic properties and magnetic properties of the carbon atoms by altering the structure of the C-C bonds in the localized regions of the material [29–34], which ultimately causes the strength of the material to weaken [35–37].

The prediction of new structures for graphene derivatives using first-principles approaches has attracted scientists' attention in recent years due to graphene's impressive and excellent mechanical properties. The most predicted new two-dimensional carbon materials exhibit specialized structures and excellent properties, rendering them highly promising for specific engineering applications such as lithium batteries, hydrogen storage, and more [38–43]. However, there are still difficulties and challenges in preparing these new materials. To accelerate the process of designing, developing, manufacturing, and ultimately engineering these new materials, MD methods were used to test the physical characteristics of the new two-dimensional carbon materials [44–48]. The results of MD testing can be utilized to inform the development of novel materials and practical engineering applications, thereby accelerating the process of integrating new materials into industrial applications.

PHTOH-graphene is a recently predicted two-dimensional carbon material with an inherent low-energy structure that provides the material with good mechanical properties and thermal stability, and researchers have used the AIMD method to study its basic physical properties [49]. When studying the mechanical properties of materials on a larger scale, AIMD is no longer applicable due to its much higher computational cost compared to classical MD. Our study used the classical MD approach to investigate the mechanical stability, elastic, and fracture properties in different orientations (armchair and zigzag). It also investigated the effects of temperature, defects (random vacancy defects and type I cracks), and hydrogen atom adsorption at the crack tip on the material's mechanical properties. This research will provide a reference for the possible subsequent practical applications of PHOTH-graphene.

2. Materials and Methods

As illustrated in Figure 1, PHOTH-graphene comprises four-, five-, six-, seven-, and eight-membered carbon rings with ten carbon atoms in its protocells. The lattice constants a and b are labeled in Figure 1, and the C-C bonds exhibit bond lengths between 0.1410 nm and 0.1435 nm. The lattice is orthogonally aligned in the P1 space group. In the model, the X and Y directions are defined in terms of the orientation of the six-membered ring as the armchair and zigzag directions. The model utilized in this study comprises 24,800 carbon atoms, and the model's size in the plane is 27.62×27.48 nm. Concerning the thickness of single-layer graphene, the thickness of single-layer PHOTH-graphene is set to 0.334 nm. The simulated box has a thickness of 13.74 nm in the z-direction, with single-layer PHOTH-graphene situated in the middle of the box in the z-direction. The impact of the model's size on the simulation will be discussed in the subsequent section.

This study conducted simulations using the classical molecular dynamics method, and the simulation software selected was LAMMPS (lammps-23Jun2022) [50]. The visualization software utilizes OVITO (Version 3.7.10) [51]. The basic parameters of the simulation included a time step of 0.0001 ps and a strain rate of 0.001 ps^{-1}; the temperature was maintained at 300 K throughout the simulation, except during the investigation of temperature effects.

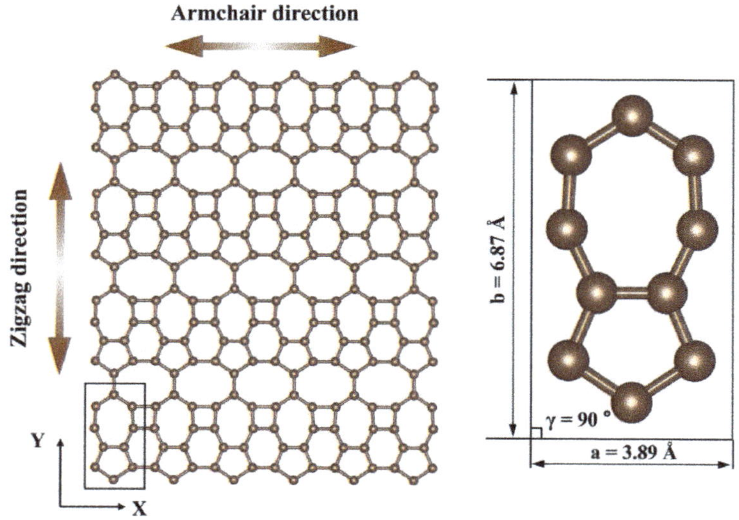

Figure 1. Schematic structure of PHOTH-graphene (**left**), and a primitive unit cell (**right**) with lattice constants values. All atoms are coplanar.

The stress–strain relationship is considered the most significant indicator during in-plane mechanical property testing of two-dimensional carbon materials and is commonly used to evaluate the mechanical properties of two-dimensional materials [22,45,52,53]. The stress–strain curve can be used to extract some crucial pieces of information. Young's modulus is calculated by fitting a linear function to the initial segment of the stress–strain curve. Ultimate stress and strain correspond to the points of maximum stress and strain, respectively, where failure typically occurs. Toughness is determined by integrating the area under the stress–strain curve.

The influence of defects on the fracture failure of a material is critical. It is commonly observed that failure of a material occurs preferentially at locations in the material where defects are present due to the stress concentrations that occur at these locations under load [22,45,54]. Two forms of defects are considered in this paper: random vacancy defects and type I cracks. The former is achieved by random deletion of carbon atoms, while the latter is achieved by deletion of carbon atoms in a designated region to create a certain crack shape. Type I cracks, also known as tensile cracks, are characterized by the dominance of normal tensile stress on the crack surface. The crack extension is oriented perpendicular to the applied stress. Griffith's theory is frequently employed to describe the fracture behavior of two-dimensional carbon materials in the presence of pre-cracks. Its mathematical form is as follows [55]:

$$\sigma_f = \sqrt{\frac{2E\gamma_s}{\pi a}} \tag{1}$$

where γ_s—surface free energy; E—Young's modulus; a—crack length; and σ_f—fracture stress.

The fracture stress of a material is closely associated with the size of the crack. Materials with cracks typically exhibit expansion from the crack's tip, which ultimately leads to the material's fracture failure.

The 2nd generation REBO potential function is used to define interactions between carbon atoms [56] and is widely employed to study the physical behavior of two-dimensional carbon materials [16,57,58]. We used an optimized REBO potential function suitable for PHOTH-graphene. The details of the optimization will be discussed in the next section.

The mathematical form of the REBO potential function, which describes the interaction between the C-C bond, is presented in Equation (2):

$$E_{ij}^{Rebo} = f(r_{ij})\left[V_{ij}^{R}(r_{ij}) + b_{ij}V_{ij}^{A}(r_{ij})\right] \quad (2)$$

where i and j indicate the serial numbers of two different atoms, respectively, r_{ij}—the distance between two atoms, b_{ij}—the bond order term that regulates the bond strength based on the bonding environment, V_{ij}^{R}—mutual repulsive interactions between atoms, V_{ij}^{A}—mutual attraction between atoms, and $f(r_{ij})$—cut-off function.

The cut-off function is shown in Equation (3):

$$f(r_{ij}) = \begin{cases} 1, & r_{ij} < rcmin\_CC \\ \frac{1+\cos\left[\pi\left(r_{ij}-rcmin\_CC\right)/(rcmax\_CC-rcmin\_CC)\right]}{2}, & rcmin\_CC < r_{ij} < rcmax\_CC \\ 0, & r_{ij} > rcmax\_CC \end{cases} \quad (3)$$

Among them, rcmin_CC and rcmax_CC are the cut-off radius of the soft cut-off, and they define the change in the cut-off function from 0 to 1. Obviously, as the interatomic distance r_{ij} increases, the value of the cut-off function gradually increases from 0 to 1, which avoids an abrupt change in the cut-off function in the numerical value, and this method of taking the value of the cut-off radius is known as the soft cut-off. Material fracture is an abrupt process, and the cut-off function is set to avoid the computational problems that abrupt changes may cause. Such a practice exists in the Tersoff, REBO, and AIREBO potential functions. rcmin_CC = 1.70 and rcmax_CC = 2.0 are the defaults in the REBO potential function, and the introduction of a soft truncation makes the fracture process of the C-C bond smoother, which is helpful in capturing the physical response of the material during the fracture process. However, a soft cut-off that interferes excessively with the fracture behavior of the material results in a material that exhibits non-physical strain hardening.

To overcome the non-physical strain hardening observed in the stress–strain curve due to the cut-off radius, a hard cut-off (r = rcmin_CC = rcmax_CC = 2.0) with a cut-off radius of 2.0 has been proposed for use in the simulation of the fracture failure of 2D graphene. This approach allows for simplified calculations while avoiding the effect of an excessive soft cut-off on the fracture behavior of the material, as the C-C bond will rapidly undergo brittle fracture after reaching the ultimate strain. This methodology avoids the excessive influence of soft cut-off on the fracture behavior of the material, thereby resulting in a calculation that is more in line with the experimental results [59–64]. The cut-off function for a hard cut-off is shown in Equation (4):

$$f(r_{ij}) = \begin{cases} 1, & r_{ij} < r \\ 0, & r_{ij} > r \end{cases} \quad (4)$$

where r_{ij}—distance between atoms and r—cut-off radius.

Furthermore, while maintaining the benefit of the soft cut-off of the REBO potential function, the potential function is optimized by adjusting the value of rcmin_CC (the typical values are 1.92 and 1.95). Additionally, the optimized potential function also demonstrates satisfactory outcomes [17,65]. In summary, the current solution to strain hardening of materials exhibiting non-physical behavior mainly consists of setting a hard cut-off and adjusting the cut-off radius of the soft cut-off.

Functional adsorption on two-dimensional carbon materials has a critical effect on fracture failure. The adsorption of hydrogen atoms on graphene, in particular, has been shown to result in a reduction in the mechanical properties of these materials [27,66,67]. Two-dimensional carbon materials have good prospects for applications in the direction of hydrogen storage, so a hydrogen-filled atmosphere would be a potential working environment for PHOTH-graphene. Building on the analysis of microcracks' impact on the material's fracture failure, we further investigated the influence of hydrogen atom adsorption at the crack tip on the material's fracture failure.

3. Results and Discussion

3.1. Optimization of the Potential Function

The uniaxial stretching of PHOTH-graphene was carried out using the original REBO potential function in order to test its mechanical properties, and the stress–strain curves exhibited a non-physical behavior of material hardening. Based on the analysis above, we believe that this is due to excessive interference by the soft cut-off on the fracture behavior of the material. In previous studies, the convergence test on the value of the cut-off radius has been demonstrated to be an essential method for optimizing the potential function, which allows for an assessment of the value of the cut-off radius, which in turn enables the selection of a more appropriate cut-off radius. Furthermore, the optimized potential function can be employed in uniaxial tensile testing to enhance the ability to capture the mechanical response of the material behavior [68,69]. Therefore, it is necessary to modify the cut-off radius of the soft cut-off in order to use it to test the mechanical properties of PHOTH-graphene.

As shown in Figure 2, we tested the convergence of the cut-off radius of the soft cut-off by keeping the value of rcmax_CC unchanged at 2.0 Å and taking rcmin_CC at 1.60 Å, 1.65 Å, 1.70 Å, 1.75 Å, 1.80 Å, 1.85 Å, 1.90 Å, and 1.95 Åy. Figure 2a shows that the stress–strain curves of uniaxial stretching exhibits the non-physical behavior of strain-hardening during the stretching process for all values of rcmin_CC that were less than 1.80 Å. As shown in Figure 2b, the ultimate stress and strain converge with the gradual increase in rcmin_CC to 1.80 Å. The ultimate stresses at rcmin_CC values of 1.60 Å, 1.65 Å, 1.70 Å, and 1.75 Å increase by 34.58%, 31.18%, 26.89%, and 14.46%, respectively, compared to the ultimate stresses at rcmin_CC of 1.80 Å. The ultimate strains at 1.85 Å, 1.90 Å, and 1.95 Å decreased by 2.48%, 4.43%, and 5.21%, respectively, compared to the ultimate stresses for rcmin_CC at 1.80 Å. Figure 2c demonstrates the variation in Young's modulus with the cut-off radius. The Young's modulus varies more for rcmin_CC values less than 1.80 Å. For the rcmin_CC values of 1.85 Å, 1.90 Å, and 1.95 Å, the Young's modulus decreases by 1.89%, 2.78%, and 3.13% compared to the Young's modulus when rcmin_CC is 1.80 Å. The Young's modulus tends to converge for values of rcmin_CC greater than 1.80 Å (values of rcmin_CC equal 1.85 Å, 1.90 Å, and 1.95 Å) and decreases by 1.89%, 2.78%, and 3.13%, respectively. Figure 2d demonstrates the variation in toughness with the change in cut-off radius; the toughness varies more for rcmin_CC values less than 1.80 Å, and for rcmin_CC values of 1.85 Å, 1.90 Å, and 1.95 Å, the toughness increases by 2.97%, 2.01%, and 5.51% compared to the toughness when rcmin_CC is 1.80 Å. The toughness of the material tends to converge after the value of rcmin_CC is greater than 1.80 Å. The toughness of the material increases by 2.97%, 2.01%, and 5.51% when the value of rcmin_CC is greater than 1.80 Å. In summary, the mechanical properties of the material begin to converge after rcmin_CC is taken to be 1.80 Å. Therefore, rcmin_CC = 1.80 Å is taken as the soft cut-off.

Furthermore, we also tested the applicability of the hard cut-off (rcmin_CC = rcmax_CC = 2.0 Å) for the mechanical property testing of PHOTH-graphene. Figure 3 shows the system's potential energy change during the relaxation process. The potential energy of the system grows significantly. The structure changes significantly under the hard cut-off, whereas the soft cut-off with rcmin_CC = 1.80 Å is used for the system's structural integrity and the potential energy of the system remains well stabilized during the relaxation process. Therefore, the hard cut-off is not suitable for testing the physical behavior of the new PHOTH-graphene material. Combining our tests on the convergence of the cut-off radius in the soft cut-off approach and the feasibility of the hard cut-off approach, we finally conclude that the soft cut-off approach is more suitable for testing the mechanical properties of PHOTH-graphene. The optimized REBO potential function with rcmin_CC = 1.80 Å and rcmax_CC = 2.0 Å was used in the subsequent studies.

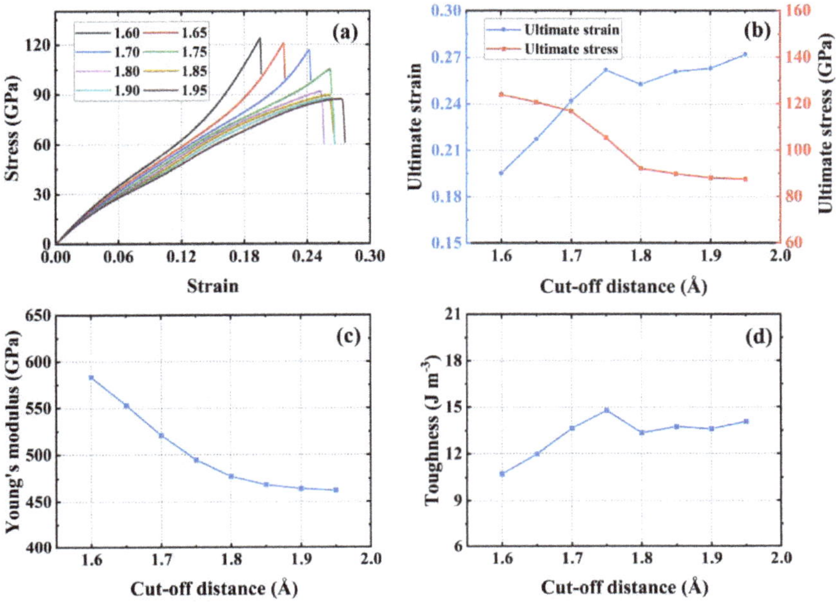

Figure 2. Convergence test for cut-off radius. (**a**) Stress–strain curves; (**b**) ultimate stress and strain; (**c**) Young's modulus; (**d**) toughness.

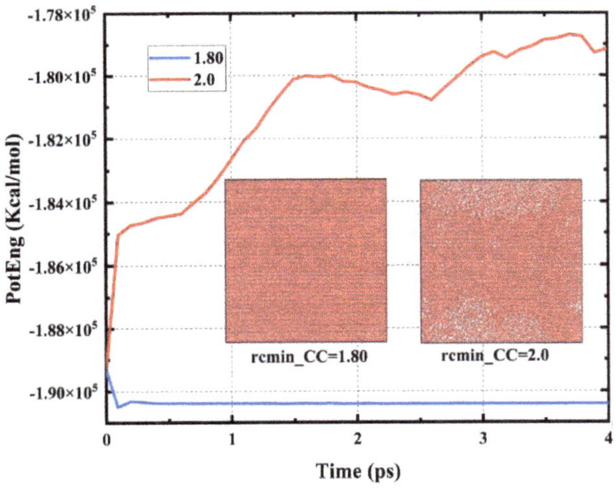

Figure 3. Potential energy changes in the simulated system during the relaxation process using both the hard cut-off and soft cut-off methods.

3.2. Size Effects

Selecting an appropriate system size can help to obtain more accurate results more economically. In general, it is difficult to obtain accurate results with too small a system of the model, and a larger system helps to obtain more accurate results. However, simply increasing the system size of the simulation shows limited improvements in the accuracy of the simulation results, so it is necessary to perform a convergence test on the system size of the simulation. As shown in Figure 4, the system size was tested for convergence using systems containing 2200, 4480, 8580, 15,900, 28,400, and 44,000 carbon atoms, respectively, along the armchair direction.

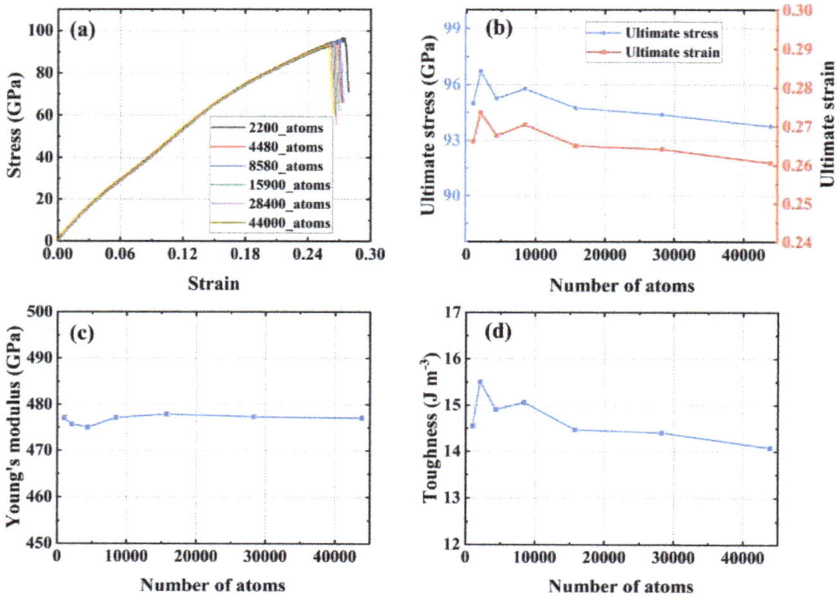

Figure 4. System size convergence tests. (**a**) Stress–strain curve; (**b**) ultimate stress and strain; (**c**) Young's modulus; (**d**) toughness.

Figure 4a illustrates the stress–strain relationships from the uniaxial tensile tests for six different sized systems. Figure 4b shows the ultimate stress and strain obtained from testing systems of different sizes. As the system size increases, the values of the ultimate stress and strain tend to stabilize. The system containing 28,400 carbon atoms shows a change of −0.49% in ultimate stress and −1.99% in ultimate strain compared to the system with 15,900 carbon atoms. Figure 4c shows the Young's modulus from uniaxial tensile testing of systems with different sizes. The Young's modulus for the four systems, each with a number of carbon atoms greater than or equal to 8580, remains essentially the same, with a maximum difference of only 2.01 GPa. Figure 4d demonstrates the toughness of the material used in the uniaxial tensile tests conducted for systems of different sizes, and the toughness measured for the three systems with a number of carbon atoms greater than or equal to 15,900 tends to converge, with a difference of only 0.39 J m^{-3} between the maximum and minimum values. To guarantee the precision and efficacy of the calculations, we decided that subsequent studies in this paper would contain 28,400 carbon atoms in the system used for analysis.

3.3. Mechanical Properties

3.3.1. Mechanical Properties in Different Orientations

The mechanical properties of PHOTH-graphene are listed in Table 1. The Young's modulus is 26.14% higher in the zigzag direction compared to the armchair direction. The ultimate stress and strain in the armchair direction are also higher than in the zigzag direction by 29.04% and 25.12%, respectively. The material's toughness is 36.46% greater in the armchair direction than in the zigzag direction. In conclusion, the material shows excellent mechanical properties in both directions.

Table 1. Mechanical properties of PHOTH-graphene in the zigzag and armchair directions.

Tensile Direction	Ultimate Stress (GPa)	Ultimate Strain	Young's Modulus (GPa)	Toughness ($J\ m^{-3}$)
Along zigzag	73.14	0.211	602.16	10.56
Along armchair	94.38	0.264	477.38	14.41

Furthermore, we conducted a comparative analysis of PHOTH-graphene and graphene. The simulation of graphene was performed using the AIREBO potential function with a hard truncation, with a cut-off radius of 2.0 Å. The remaining simulation conditions remained consistent with those of PHOTH-graphene. Figure 5 shows that Young's modulus of graphene is 964.07 GPa and 1007.99 GPa in the zigzag and armchair directions, which is consistent with earlier studies [9,69] and is considerably higher than that of PHOTH-graphene in both directions. The exceptional mechanical properties of graphene are attributed to its distinctive six-membered ring structure. To elucidate the rationale behind the disparity in the mechanical properties of graphene and PHOTH-graphene, we introduce the concept of mass surface density, which reflects the structural characteristics of two-dimensional carbon materials in the plane to a certain extent. It is commonly observed that the differing structural characteristics of two-dimensional carbon materials will result in disparate mass surface density.

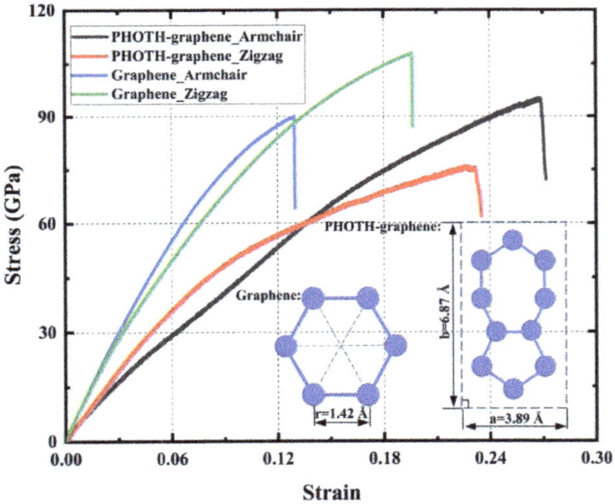

Figure 5. Stress–strain curves of PHOTH-graphene and graphene in different tensile directions.

The mass surface density of the two-dimensional carbon material can be calculated using Equation (5).

$$\rho_{ij} = \frac{m}{S} \tag{5}$$

where m—the total mass of carbon atoms and S—the area of the region containing a certain number of carbon atoms.

The mass face density of graphene is calculated using Equation (5) as 7.66×10^{-7} kg/m^2, while that of PHOTH-graphene is calculated as 7.46×10^{-7} kg/m^2. The face density of graphene is found to be 2.7% greater than that of PHOTH-graphene, resulting in graphene exhibiting superior mechanical properties and a larger Young's modulus. Mass face density clearly explains the difference in the mechanical properties between graphene and PHOTH-graphene.

The projected line density of a two-dimensional carbon material can be calculated from the surface density.

$$\begin{cases} \rho_i = \rho_{ij} \times l_j \\ \rho_j = \rho_{ij} \times l_i \end{cases} \quad (6)$$

where ρ_{ij}—mass surface density, ρ_i—projected line density of the material along the i-direction, ρ_j—projected line density of the material along the j-direction, l_i—the length of the projection of the region along the i-direction, and l_j—the length of the projection of the region along the j-direction.

The projected linear densities of PHOTH-graphene along the armchair and zigzag directions can be calculated from Equation (6) as 5.125×10^{-16} kg/m and 2.902×10^{-16} kg/m. The ultimate stress and strain along the armchair direction in the stress–strain curves are larger, which show stronger tensile toughness. However, the Young's modulus along the armchair direction is smaller than that along the zigzag direction. This difference arises because the projected line density, unlike the mass surface density, is one-dimensional and overlooks the material's structural characteristics along the projected direction. Since Young's modulus is closely related to the material's structure, this oversight leads to a lower modulus in the armchair direction.

3.3.2. Forms of Failure and Fracture

To gain further insight into the failure and fracture behavior of PHOTH-graphene, we took snapshots of the material during uniaxial stretching in both directions. As shown in Figure 6, the material shows microcracks at a strain of 0.254 in the armchair direction. By the time the strain stage reaches 0.256, the crack size has expanded rapidly, and a region containing an amorphous structure has appeared in the direction of the crack tip. As the strain increases to 0.258, the crack size expands further, and the area containing the amorphous structure increases. Eventually, the material fails and fractures when the strain reaches the ultimate strain of 0.264.

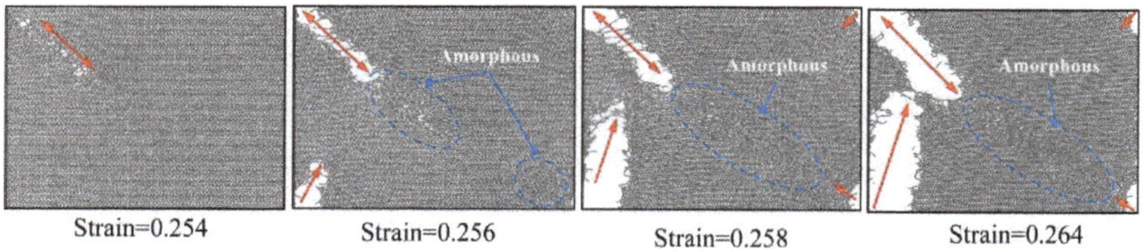

Figure 6. Failure fracture behavior of PHOTH-graphene under uniaxial tension along the armchair direction.

As shown in Figure 7, the material did not show significant microcracking and phase transformation when the material reached the ultimate strain of 0.211 in the zigzag direction. Further increasing the strain to 0.220 on the basis of the ultimate strain, the material shows obvious regions containing amorphous structures. At a strain of 0.224, the area of the material containing amorphous structure increases. At a strain of 0.300, the area of the material containing the amorphous structure increases further and microcracks appear in the amorphous structure. At a strain of 0.303, the size of the cracks is rapidly expanded. At a strain of 0.305, further expansion of the crack size occurs.

The form of failure fracture differs between the armchair and zigzag directions. Microcracks are first observed during uniaxial stretching in the armchair direction, but the material does not fail entirely at this stage. As the crack expands towards the crack tip, it encounters a region with an amorphous structure. The material then reaches ultimate strain with rapid crack expansion, eventually leading to brittle fracture failure. A region containing an amorphous structure is first observed during uniaxial stretching in the zigzag direction. The appearance of the amorphous structure significantly weakens the material's overall bearing capacity, indicating initial failure. Microcracks then appear near the amorphous region, and the cracks expand rapidly with further strain increase.

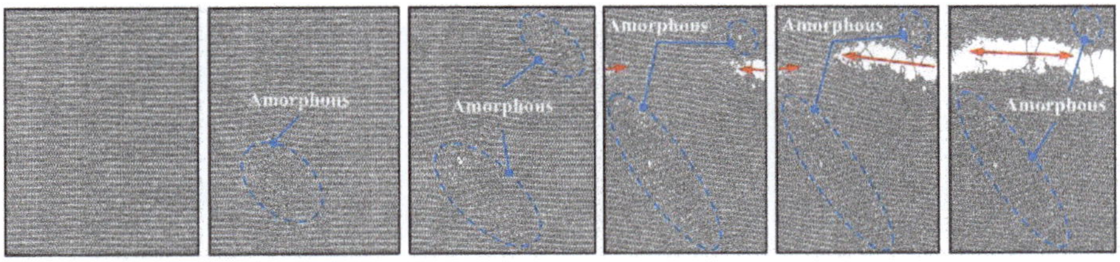

Figure 7. Failure fracture behavior of PHOTH-graphene under uniaxial tension along the zigzag direction.

3.3.3. Mechanical Properties at Different Temperatures

To investigate the influence of temperature on the mechanical properties of PHOTH-graphene, we tested the material's properties along the armchair and zigzag directions at temperatures of 200 K, 300 K, 400 K, 500 K, 600 K, 700 K, 900 K, and 1200 K.

Figure 8a shows the stress–strain curves of PHOTH-graphene in the armchair direction at different temperatures. The variation in ultimate stress and strain with temperature changes is illustrated in Figure 8b. As the temperature increases from 200 K to 1200 K, the ultimate stress and strain decrease by 56.69% and 57.76%, respectively. Figure 8c presents Young's modulus at different temperatures, with values of 479.75 GPa at 200 K and 444.99 GPa at 1200 K, reflecting a decrease of 7.25%. Figure 8d shows the toughness of PHOTH-graphene at different temperatures, where an increase from 200 K to 1200 K results in a decrease in toughness from 17.45 J m^{-3} to 3.24 J m^{-3}, representing an 81.43% reduction.

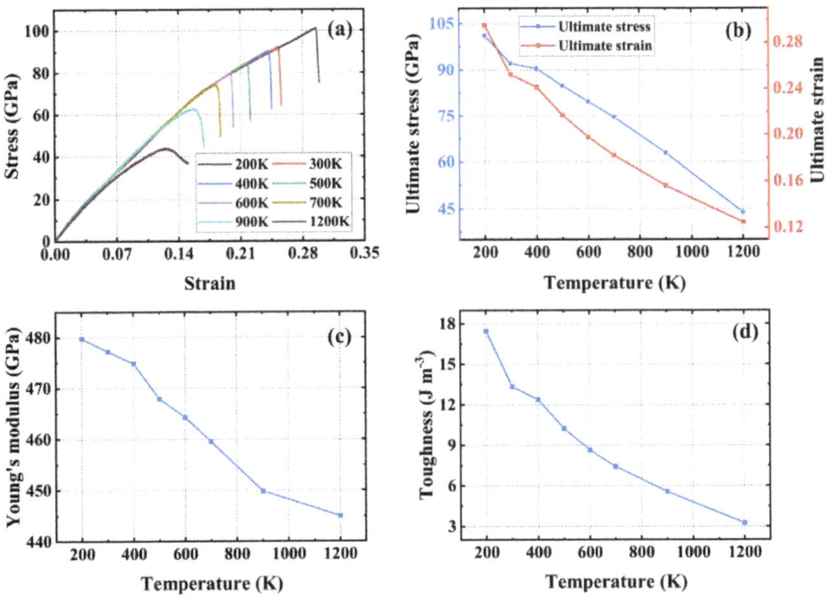

Figure 8. Mechanical properties of PHOTH-graphene in the armchair direction at different temperatures. (**a**) Stress–strain curve; (**b**) ultimate stress and strain; (**c**) Young's modulus; (**d**) toughness.

Figure 9a illustrates the stress–strain curves of PHOTH-graphene in the zigzag direction at different temperatures, showing that elevated temperature weakens the material's strength. Figure 9b shows the ultimate stress and strain at different temperatures: 77.85 GPa and 0.24 at 200 K, and 27.33 GPa and 0.06 at 1200 K. Figure 9c presents the Young's modulus at different temperatures: 611.52 GPa at 200 K and 528.15 GPa at 1200 K. Figure 9d shows the toughness of the material at different temperatures: 12.32 J m^{-3} at 200 K and 1.01 J m^{-3} at 1200 K. The increase in temperature from 200 K to 1200 K results in decreases in Young's modulus, toughness, ultimate stress, and strain by 13.63%, 91.80%, 64.89%, and 75.00%, respectively.

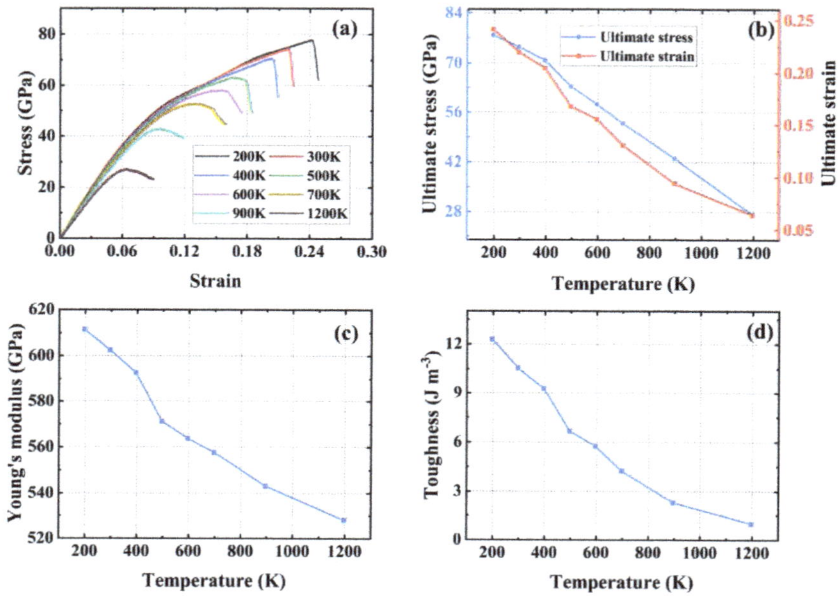

Figure 9. Mechanical properties of PHOTH-graphene in the zigzag direction at different temperatures. (**a**) Stress–strain curve; (**b**) ultimate stress and strain; (**c**) Young's modulus; (**d**) toughness.

Figure 10 shows the percentage decrease in the mechanical properties of PHOTH-graphene in both directions after increasing the temperature by 1000 K (from 200 K to 1200 K). The comparison indicates that the mechanical properties of PHOTH-graphene in the zigzag direction are more sensitive to temperature. Young's modulus is significantly less sensitive to temperature than toughness, and ultimate stress and strain.

The C-C bond is stretched until it breaks as a stress-activated process [60]. The relationship between fracture strength, strain rate, and temperature can be described by the Arrhenius equation [15]:

$$\varepsilon = A\sigma^{\frac{1}{m}}\exp(-\frac{Q}{RT}) \tag{7}$$

where ε—strain rate, A—indexing factor, m—sensitive factor, R—general gas constant, T—temperature, Q—activation energy, and σ—breaking strength.

Equation (7) shows that keeping the strain rate constant during uniaxial stretching decreases fracture strength at elevated temperatures. The higher temperature increases the vibration and kinetic energy of the atoms, which leads to more drastic changes in the bond lengths of the atomic bonds, weakened interatomic interaction forces, and a softer material that becomes more susceptible to deformation, which ultimately causes a decrease in the strength of the material [70,71].

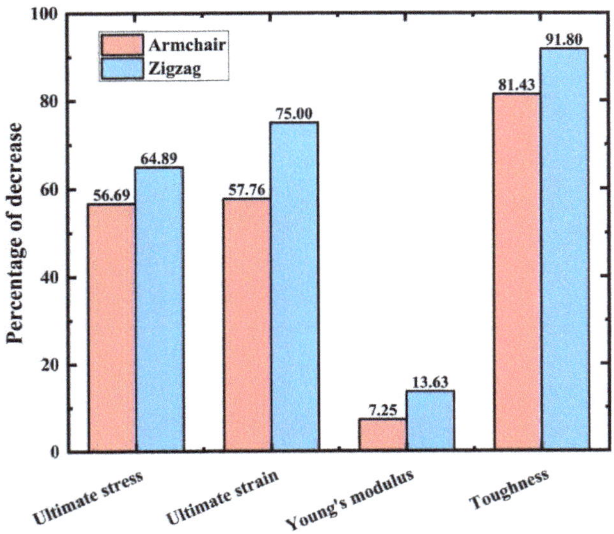

Figure 10. Percentage decrease in ultimate stress, ultimate strain, Young's modulus, and toughness after temperature increases from 200 K to 1200 K.

3.4. *Effect of Defects*

3.4.1. Random Vacancy Defects

Two-dimensional carbon materials inevitably develop defects during manufacturing and engineering applications, with vacancy defects being the most common. To explore the influence of random vacancy defects, we conducted in-plane mechanical property tests on materials without defects (0%) and with five concentrations of random vacancy defects: 1%, 2%, 3%, 4%, and 5%. Each concentration was tested three times. The PHOTH-graphene system remained stable after introducing the defects.

Figure 11a shows the stress–strain curves of the material with different concentrations of random vacancy defects in the X (armchair) direction. The addition of these defects reduces the strength of the material. Figure 11b demonstrates a monotonic decrease in Young's modulus with increasing concentration of vacancy defects. Figure 11c demonstrates the monotonically decreasing trend of the ultimate stress with increasing concentration of vacancy defects. It is noteworthy that the ultimate strain of the material initially decreases and then increases as the concentration of vacancy defects rises. As shown in Figure 11d, the ultimate strain of the material monotonically decreases and the ultimate strain decreases by 19.86% when the vacancy defect concentration increases from 0% to 3%, while the ultimate strain of the material monotonically increases and the ultimate strain increases by 8.25% when the vacancy defect concentration increases from 3% to 5%. In summary, the mechanical properties of PHOTH-graphene in the armchair direction show a decrease in the ultimate stress, ultimate strain, and Young's modulus of the material with the addition of different proportions of random vacancy defects.

Figure 12a illustrates the stress–strain curves for uniaxial stretching with the addition of different vacancy defect concentrations in the Y (zigzag) direction. Figure 12b,c demonstrate the monotonic decrease in the Young's modulus and ultimate stress with an increase in the vacancy defect concentration, respectively. Figure 12d demonstrates the trend of the ultimate strain of the material with an increasing concentration of vacancy defects; compared to the material without added vacancy defects, the ultimate strain of the material with 1% and 2% concentration of vacancy defects decreases by 8.06% and 1.89%. The ultimate strain of the material with 3%, 4%, and 5% concentration of vacancy defects increases by 8.53%, 12.32%, and 22.27%.

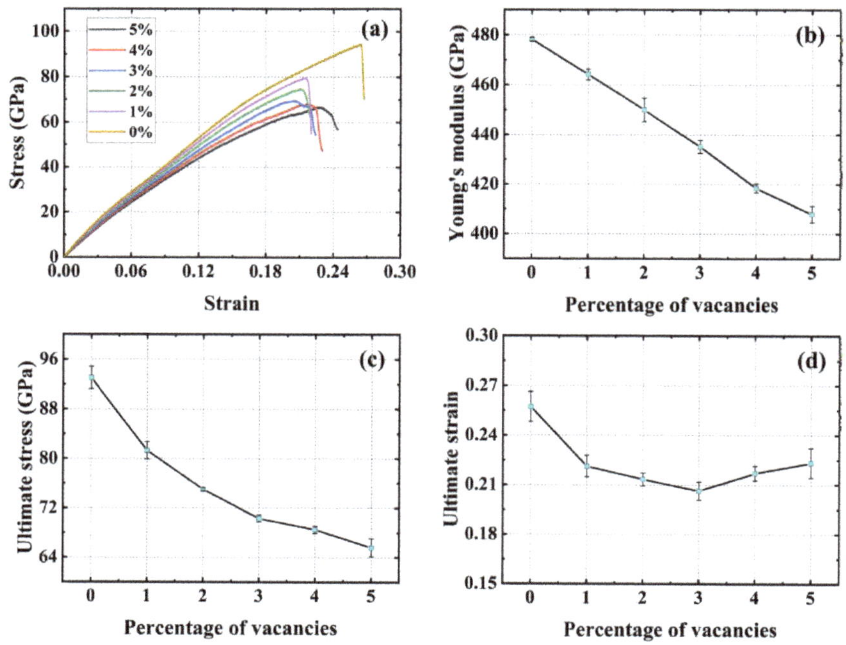

Figure 11. Mechanical properties of PHOTH-graphene in the X (armchair) direction for different random vacancy defect concentrations. (**a**) Stress–strain curve; (**b**) Young's modulus; (**c**) ultimate stress; (**d**) ultimate strain.

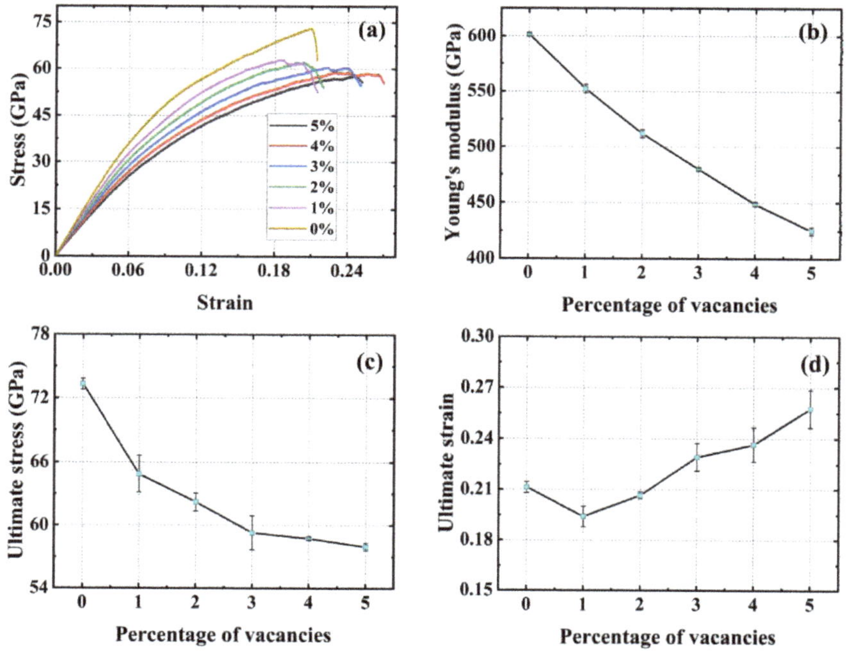

Figure 12. Mechanical properties of PHOTH-graphene in the Y (zigzag) direction for different random vacancy defect concentrations. (**a**) Stress–strain curve; (**b**) Young's modulus; (**c**) ultimate stress; (**d**) ultimate strain.

It is noteworthy that the ultimate strain of the material did not decrease monotonically with increasing percentage of random vacancy defects, and Figure 13 further compares the changes in the ultimate strain of PHOTH-graphene in the armchair and zigzag directions with increasing percentage of random vacancy defects. In the armchair direction, the addition of random vacancy defects causes the ultimate strain of the material to decrease, and the percentage of random vacancy defects after 3% of the ultimate strain no longer decreases but rather increases again. In the zigzag direction, the addition of 1% and 2% of random vacancy defects causes the ultimate strain of the material to decrease, and further increasing the percentage of random vacancy defects to 3%, 4%, and 5% increases the ultimate strain, which improves the strength of PHOTH-graphene to some extent. At random vacancy defect concentrations below 3%, the ultimate strain in both directions was reduced for PHOTH-graphene with added random vacancy defects compared to the material without added defects. However, at random vacancy defect concentrations higher than 3%, the ultimate strain in the zigzag direction was instead greater for the material with added defects than for the material without added defects.

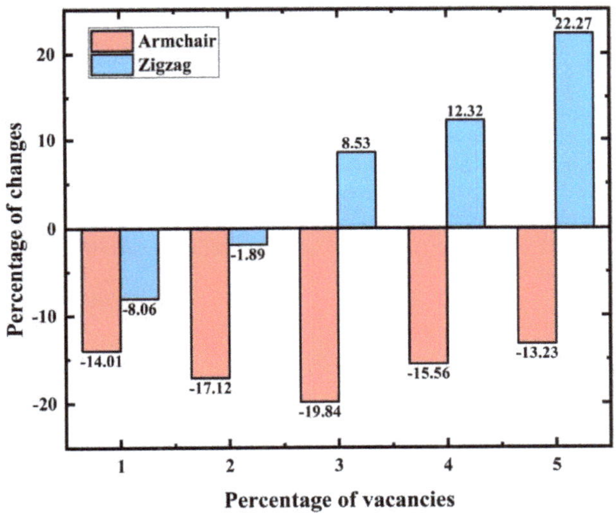

Figure 13. Percentage change in ultimate strain with different percentages of random vacancy defects. The percentage of random vacancy defects increased from 1% to 5%.

3.4.2. Type I Cracks and Crack Tip Hydrogen Absorption

Cracks affect the mechanical properties of materials, with type I cracks being the most common form of cracks causing fracture failure of materials. As shown in Figure 14a, in this paper, type I cracks with lengths of 10 Å, 23 Å, and 51 Å are further considered on the basis of a model without defects. The width of the crack is 8 Å. The model remains stable during model relaxation, and the edge structure of the crack does not change.

As shown in Figure 14b, to further explore the influence of hydrogen adsorption at the crack tip on the material's strength, we considered three different crack tips with a length of 51 Å—0H, 4H, and 8H, indicating 0, 4, and 8 hydrogen atoms adsorbed at the crack tip, respectively. The direction of uniaxial stretching is labeled in the figure. It should be noted that the adsorption of hydrogen atoms at the crack tip is all in-plane adsorption. This study preliminarily investigates the effect of hydrogen atom adsorption at the crack tip without exploring all possible adsorption sites or numbers of hydrogen atoms.

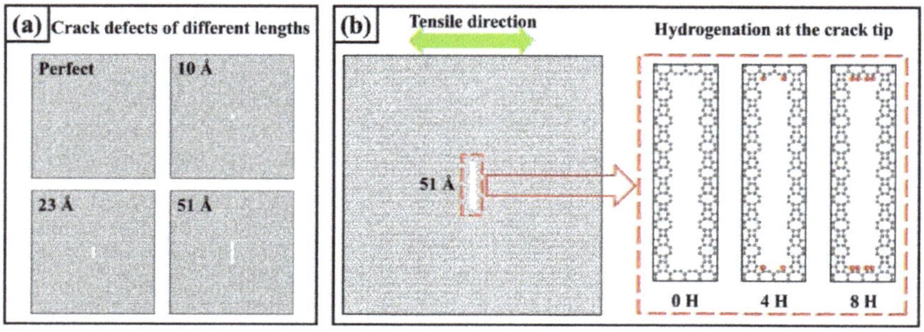

Figure 14. Models for hydrogen absorption at the crack tip and cracks. (**a**) Models for crack defects with different lengths; (**b**) Models for hydrogen absorption at the crack tip.

Figure 15 shows that cracks of different lengths weaken the strength of PHOTH-graphene to varying degrees. Compared with the model without cracks, the cracks with lengths of 10 Å, 23 Å, and 51 Å reduced the Young's modulus by 0.15%, 1.24%, and 3.46%; the ultimate stress by 9.48%, 23.84%, and 42.16%; and the ultimate strain by 14.77%, 28%, and 28%, respectively. The ultimate stresses were reduced by 9.48%, 23.84%, and 42.16%; the ultimate strains were reduced by 14.77%, 28.42%, and 49.24%; and the toughness was reduced by 25.12%, 46.43% and 73.07%, respectively. Young's modulus is obviously less sensitive to crack length changes than ultimate stress, strain, and toughness. Figure 17a shows the von Mises stress distribution during uniaxial stretching of a material with a crack length of 51 Å. An apparent stress concentration occurs at the crack tip, causing the material to fail preferentially at this location. Further extension of the crack ultimately leads to the material's fracture failure.

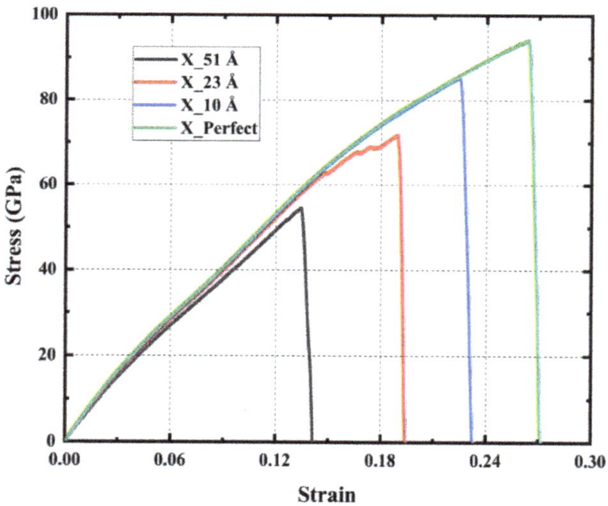

Figure 15. Stress–strain curves of models with different length cracks under uniaxial tensile action.

Figure 16 shows the stress–strain curves under uniaxial tensile action for models with different crack tips, each with a crack length of 51 Å. The adsorption of hydrogen atoms at the crack tip somewhat reduces the material's mechanical properties. Compared to crack tips with no hydrogen atoms, crack tips with four and eight hydrogen atoms result in reductions in Young's modulus by 0.22% and 0.30%, ultimate stress by 6.67% and 10.46%, ultimate strain by 2.99% and 1.49%, and toughness by 6.19% and 5.93%, respectively. The value of Young's modulus is insensitive to hydrogen adsorption at the crack tip and hardly changes after adsorption.

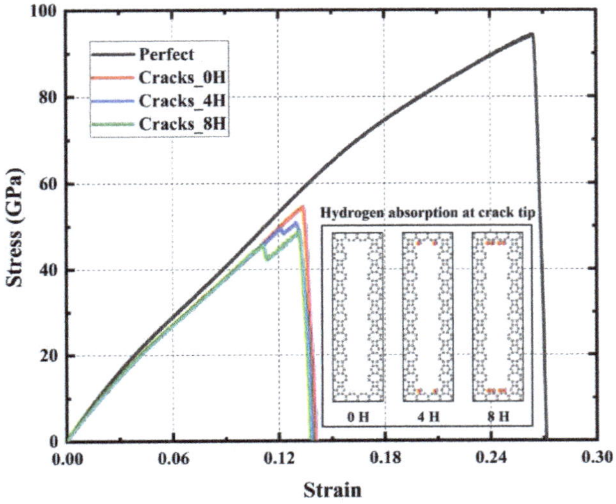

Figure 16. Stress–strain curves of models with different crack tips under uniaxial tensile action.

Table 2 lists the values of the mechanical properties under uniaxial tension for a perfect material without crack defects, a material with three different crack sizes, and a model with three different crack tips.

Table 2. Mechanical properties of PHOTH-graphene with cracks.

Type of Crack	Ultimate Stress (GPa)	Ultimate Strain	Young's Modulus (GPa)	Toughness (J m^{-3})
Perfect	94.38	0.264	477.38	14.41
10 Å	85.43	0.225	476.65	10.79
23 Å	71.88	0.189	470.39	7.72
51 Å	54.59	0.134	461.88	3.88
51 Å_4H	50.95	0.130	460.87	3.64
51 Å_8H	48.88	0.132	460.50	3.65

Figure 17 illustrates the von Mises stress distribution during fracture for cracks with three different crack tips, all exhibiting significant stress concentrations at the crack tip. At a strain of 0.135, the model crack with hydrogen atoms adsorbed at the crack tip expands rapidly, while the crack with no hydrogen atoms begins to form. Hydrogen atoms cause the crack to expand at a much smaller strain because the strength of the C-C bonds at the hydrogen adsorption sites is weakened [33,36,66,67,72]. Thus, hydrogen atom adsorption at the crack tip further weakens the material's mechanical properties.

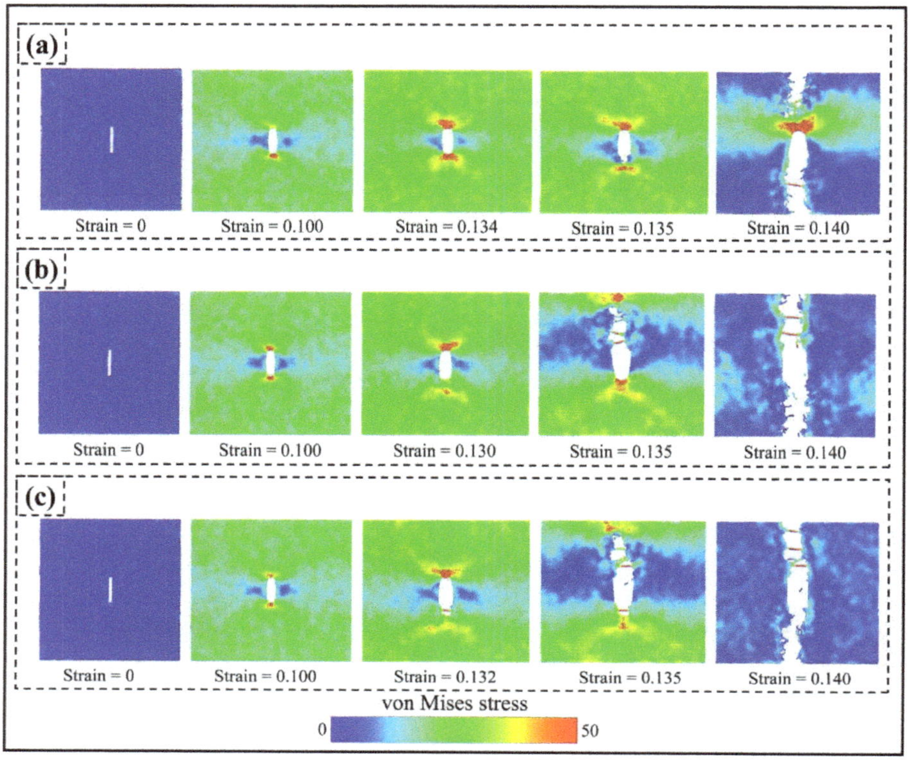

Figure 17. Von Mises stresses during fracture failure for models with three different crack tips: (**a**) no hydrogen atoms adsorbed at the crack tip; (**b**) four hydrogen atoms adsorbed at the crack tip; (**c**) eight hydrogen atoms adsorbed at the crack tip. The crack lengths are all 51 Å.

4. Conclusions

This study tested the mechanical properties of a new two-dimensional carbon material, PHOTH-graphene, using an optimized REBO potential function based on a classical molecular dynamics approach. The following conclusions were obtained:

(1) PHOTH-graphene exhibits excellent mechanical properties in both the armchair and zigzag directions, but there are also significant differences. The material has stronger ultimate stress, ultimate strain, and toughness in the armchair direction and greater Young's modulus in the zigzag direction.

(2) The failure fracture patterns of PHOTH-graphene differ between the armchair and zigzag directions. Microcracks are first observed when the material is stretched uniaxially in the armchair direction, and regions of the amorphous structure appear in the direction of crack tip expansion. Finally, brittle fracture occurs as the crack expands. The region containing the amorphous structure is observed first when the material is stretched uniaxially in the zigzag direction; then, microcracks appear in the neighborhood of the amorphous region; and the material fails due to fracture as the crack expands.

(3) Increasing the temperature weakens the strength of PHOTH-graphene. An increase in temperature by 1000 K (from 200 K to 1200 K) results in varying temperature sensitivities of the material's mechanical properties in both directions. Young's modulus is less sensitive to temperature increases than toughness, ultimate stress, and strain.

(4) The introduction of random vacancy defects significantly weakened the strength of PHOTH-graphene. The values of ultimate stress, Young's modulus, and toughness all

decreased monotonically with increasing concentrations of random vacancy defects. Notably, increasing the concentration of random vacancy defects can enhance the ultimate strain of the material. Particularly, after increasing the concentration of vacancy defects by 3% in the zigzag direction, the ultimate strain of the material is higher than that of the material with no vacancy defects.

(5) The stress concentration area during uniaxial stretching appears at the tip of a type I crack, significantly weakening the mechanical properties of PHOTH-graphene. The adsorption of hydrogen atoms at the crack tip induces the crack to expand under more minor strains, further weakening the material's strength.

Author Contributions: Conceptualization, Q.P. and Z.H. (Zhongwei Hu); Data curation, G.C. and A.L.; Funding acquisition, Q.P., X.-J.C. and Z.H. (Zhongwei Hu); Methodology, G.C. and X.C. (Xue Chen); Project administration, Q.P. and X.-J.C.; Software, G.C. and Z.H. (Zeyu Huang); Supervision, Q.P. and Z.H. (Zhongwei Hu); Validation, X.C. (Xintian Cai), Y.Z. and G.C.; Visualization, X.C. (Xintian Cai), Y.Z. and Z.H. (Zeyu Huang); Writing—original draft, G.C.; Writing—review and editing, X.C. (Xintian Cai), X.C. (Xue Chen) and A.L. All authors have read and agreed to the published version of the manuscript.

Funding: This research was funded by the Shenzhen Science and Technology Program (Grant No. KQTD20200820113045081), the National Natural Science Foundation of China (grant No. 12272378), the Strategic Priority Research Program of the Chinese Academy of Sciences (grant No. XDB0620103), and the High-level Innovation Research Institute Program of Guangdong Province (grant No. 2020B0909010003).

Institutional Review Board Statement: Not applicable.

Informed Consent Statement: Not applicable.

Data Availability Statement: The original contributions presented in the study are included in the article, further inquiries can be directed to the corresponding authors.

Conflicts of Interest: The authors declare no conflicts of interest.

References

1. Torkaman-Asadi, M.A.; Kouchakzadeh, M.A. Atomistic simulations of mechanical properties and fracture of graphene: A review. *Comput. Mater. Sci.* **2022**, *210*, 111457. [CrossRef]
2. Gounzari, M.; Amallal, F.; Bouzelmad, M.; Belkassmi, Y.; Kotri, A. Exploring the potential of nanoporous amorphous graphene as a reverse osmosis membrane: Insights from molecular dynamics simulations. *Chem. Phys. Lett.* **2024**, *848*, 141393. [CrossRef]
3. Tao, Y.; Sui, Z.-Y.; Han, B.-H. Advanced porous graphene materials: From in-plane pore generation to energy storage applications. *J. Mater. Chem. A* **2020**, *8*, 6125–6143. [CrossRef]
4. Cohen-Tanugi, D.; Grossman, J.C. Water Desalination across Nanoporous Graphene. *Nano Lett.* **2012**, *12*, 3602–3608. [CrossRef] [PubMed]
5. Chamberland, J.P.; Sellathurai, A.J.; Scott Parent, J.; Barz, D.P.J. Optimized hydrogel electrodes for supercapacitors from high-concentration aqueous graphene nanoplatelet dispersions. *J. Power Sources* **2024**, *605*, 234435. [CrossRef]
6. Jin, X.; Feng, C.; Ponnamma, D.; Yi, Z.; Parameswaranpillai, J.; Thomas, S.; Salim, N.V. Review on exploration of graphene in the design and engineering of smart sensors, actuators and soft robotics. *Chem. Eng. J. Adv.* **2020**, *4*, 100034. [CrossRef]
7. Tan, Y.; Tao, X.; Ouyang, Y.; Peng, Q. Stable and 7.7 wt% hydrogen storage capacity of Ti decorated Irida-Graphene from first-principles calculations. *Int. J. Hydrogen Energy* **2024**, *50*, 738–748. [CrossRef]
8. Varillas, J.; Lukeš, J.; Manikas, A.; Maňák, J.; Dluhoš, J.; Melníková, Z.; Kalbáč, M.; Galiotis, C.; Frank, O. Mechanical response of monolayer graphene via a multi-probe approach. *Int. J. Mech. Sci.* **2024**, *273*, 109208. [CrossRef]
9. Lee, C.; Wei, X.; Kysar, J.W.; Hone, J. Measurement of the Elastic Properties and Intrinsic Strength of Monolayer Graphene. *Science* **2008**, *321*, 385–388. [CrossRef]
10. Liu, X.; Yu, P.; Zheng, B.; Oterkus, E.; He, X.; Lu, C. Prediction of graphene's mechanical and fracture properties via peridynamics. *Int. J. Mech. Sci.* **2024**, *266*, 108914. [CrossRef]
11. Cao, K.; Feng, S.; Han, Y.; Gao, L.; Hue Ly, T.; Xu, Z.; Lu, Y. Elastic straining of free-standing monolayer graphene. *Nat. Commun.* **2020**, *11*, 284. [CrossRef] [PubMed]
12. Robertson, A.W.; Warner, J.H. Atomic resolution imaging of graphene by transmission electron microscopy. *Nanoscale* **2013**, *5*, 4079–4093. [CrossRef] [PubMed]
13. Zhao, H.; Aluru, N.R. Temperature and strain-rate dependent fracture strength of graphene. *J. Appl. Phys.* **2010**, *108*, 064321. [CrossRef]

14. Polewczyk, F.; Leyssale, J.-M.; Lafourcade, P. Temperature-dependent elasticity of single crystalline graphite. *Comput. Mater. Sci.* **2023**, *220*, 112045. [CrossRef]
15. Zhang, Y.-Y.; Pei, Q.-X.; Mai, Y.-W.; Gu, Y.-T. Temperature and strain-rate dependent fracture strength of graphynes. *J. Phys. D Appl. Phys.* **2014**, *47*, 425301. [CrossRef]
16. Gamboa-Suárez, A.; Seuret-Hernández, H.Y.; Leyssale, J.-M. Mechanical properties of pristine and nanocrystalline graphene up to ultra-high temperatures. *Carbon Trends* **2022**, *9*, 100197. [CrossRef]
17. Grantab, R.; Shenoy, V.B.; Ruoff, R.S. Anomalous Strength Characteristics of Tilt Grain Boundaries in Graphene. *Science* **2010**, *330*, 946–948. [CrossRef]
18. Banhart, F.; Kotakoski, J.; Krasheninnikov, A.V. Structural Defects in Graphene. *ACS Nano* **2011**, *5*, 26–41. [CrossRef]
19. Rajasekaran, G.; Narayanan, P.; Parashar, A. Effect of Point and Line Defects on Mechanical and Thermal Properties of Graphene: A Review. *Crit. Rev. Solid State Mater. Sci.* **2016**, *41*, 47–71. [CrossRef]
20. Zandiatashbar, A.; Lee, G.-H.; An, S.J.; Lee, S.; Mathew, N.; Terrones, M.; Hayashi, T.; Picu, C.R.; Hone, J.; Koratkar, N. Effect of defects on the intrinsic strength and stiffness of graphene. *Nat. Commun.* **2014**, *5*, 3186. [CrossRef]
21. Gavallas, P.; Savvas, D.; Stefanou, G. Mechanical properties of graphene nanoplatelets containing random structural defects. *Mech. Mater.* **2023**, *180*, 104611. [CrossRef]
22. Bedi, D.; Sharma, S.; Tiwari, S.K. Effect of chirality and defects on tensile behavior of carbon nanotubes and graphene: Insights from molecular dynamics. *Diam. Relat. Mater.* **2022**, *121*, 108769. [CrossRef]
23. Zheng, B.; Gu, G.X. Tuning the graphene mechanical anisotropy via defect engineering. *Carbon* **2019**, *155*, 697–705. [CrossRef]
24. He, L.; Guo, S.; Lei, J.; Sha, Z.; Liu, Z. The effect of Stone–Thrower–Wales defects on mechanical properties of graphene sheets—A molecular dynamics study. *Carbon* **2014**, *75*, 124–132. [CrossRef]
25. Ajori, S.; Eftekharfar, A.R. Mechanical properties and fracture analysis of defective penta-graphene under temperature variation: Insight from molecular dynamics. *Diam. Relat. Mater.* **2022**, *124*, 108956. [CrossRef]
26. Chen, M.; Han, X.; Tang, K. Topological regulations of Stone-Wales graphene. *Carbon* **2024**, *226*, 119163. [CrossRef]
27. Peng, Q.; Liang, C.; Ji, W.; De, S. A theoretical analysis of the effect of the hydrogenation of graphene to graphane on its mechanical properties. *Phys. Chem. Chem. Phys.* **2013**, *15*, 2003–2011. [CrossRef]
28. Tozzini, V.; Pellegrini, V. Prospects for hydrogen storage in graphene. *Phys. Chem. Chem. Phys.* **2013**, *15*, 80–89. [CrossRef]
29. Pei, Q.X.; Zhang, Y.W.; Shenoy, V.B. A molecular dynamics study of the mechanical properties of hydrogen functionalized graphene. *Carbon* **2010**, *48*, 898–904. [CrossRef]
30. Boukhvalov, D.W.; Katsnelson, M.I.; Lichtenstein, A.I. Hydrogen on graphene: Electronic structure, total energy, structural distortions and magnetism from first-principles calculations. *Phys. Rev. B* **2008**, *77*, 035427. [CrossRef]
31. Balog, R.; Jørgensen, B.; Nilsson, L.; Andersen, M.; Rienks, E.; Bianchi, M.; Fanetti, M.; Lægsgaard, E.; Baraldi, A.; Lizzit, S.; et al. Bandgap opening in graphene induced by patterned hydrogen adsorption. *Nat. Mater.* **2010**, *9*, 315–319. [CrossRef] [PubMed]
32. Ebrahimi, S.; Montazeri, A.; Rafii-Tabar, H. Molecular dynamics study of a new mechanism for ripple formation on graphene nanoribbons at very low temperatures based on H_2 physisorption. *Solid State Commun.* **2013**, *159*, 84–87. [CrossRef]
33. Xie, Y.; Chen, L.; Xu, J.; Liu, W. Effective regulation of the electronic properties of a biphenylene network by hydrogenation and halogenation. *RSC Adv.* **2022**, *12*, 20088–20095. [CrossRef] [PubMed]
34. Haberer, D.; Petaccia, L.; Wang, Y.; Quian, H.; Farjam, M.; Jafari, S.A.; Sachdev, H.; Federov, A.V.; Usachov, D.; Vyalikh, D.V.; et al. Electronic properties of hydrogenated quasi-free-standing graphene. *Phys. Status Solidi (B)* **2011**, *248*, 2639–2643. [CrossRef]
35. Li, N.-N.; Sha, Z.-D.; Pei, Q.-X.; Zhang, Y.-W. Hydrogenated Grain Boundaries Control the Strength and Ductility of Polycrystalline Graphene. *J. Phys. Chem. C* **2014**, *118*, 13769–13774. [CrossRef]
36. Tabarraei, A.; Wang, X.; Jia, D. Effects of hydrogen adsorption on the fracture properties of graphene. *Comput. Mater. Sci.* **2016**, *121*, 151–158. [CrossRef]
37. Ebrahimi, S. Effect of hydrogen coverage on the buckling of penta-graphene by molecular dynamics simulation. *Mol. Simul.* **2016**, *42*, 1485–1489. [CrossRef]
38. Bhattacharya, D.; Jana, D. TPO12-graphene: A new two-dimensional metallic carbon with 4–5 ring for Lithium ion battery. *Appl. Surf. Sci.* **2024**, *669*, 160495. [CrossRef]
39. Laranjeira, J.A.S.; Martins, N.F.; Denis, P.A.; Sambrano, J.R. Graphenyldiene: A new sp2-graphene-like nanosheet. *Carbon Trends* **2024**, *14*, 100321. [CrossRef]
40. Bhattacharya, D.; Jana, D. TPDH-graphene: A new two dimensional metallic carbon with NDR behaviour of its one dimensional derivatives. *Phys. E Low-Dimens. Syst. Nanostructures* **2021**, *127*, 114569. [CrossRef]
41. Bandyopadhyay, A.; Majumdar, A.; Chowdhury, S.; Ahuja, R.; Jana, D. 8-16-4 graphyne: Square-lattice two-dimensional nodal line semimetal with a nontrivial topological Zak index. *Phys. Rev. B* **2021**, *103*, 075137. [CrossRef]
42. Bhattacharya, D.; Jana, D. Worm-graphene: A two-dimensional orthorhombic carbon semimetal with massless Dirac fermion. *Appl. Surf. Sci.* **2022**, *585*, 152457. [CrossRef]
43. Chen, H.-T.; Dhanthala Chittibabu, D.K. A new carbon allotrope: Biphenylene as promising anode materials for Li-ion and LiO_2 batteries. *Solid State Ion.* **2023**, *395*, 116214. [CrossRef]
44. Peng, Q.; Huang, Z.; Chen, G.; Zhang, Y.; Zhang, X.; Chen, X.-J.; Hu, Z. Effect of Strain Rate, Temperature, Vacancy, and Microcracks on Mechanical Properties of 8-16-4 Graphyne. *Nanomaterials* **2024**, *14*, 556. [CrossRef] [PubMed]

45. Farzin, A.; Etemadi, M.; Mehran, S.; Rouhi, S. Investigating the mechanical properties of perfect and defective Ψ-graphene: A molecular dynamics simulation. *Mater. Today Commun.* **2023**, *37*, 106908. [CrossRef]
46. Alves, R.A.F.; Giozza, W.F.; Ribeiro, L.A.; Pereira, M.L. Exploring the thermal and mechanical properties of PAI-Graphene monolayers and nanotubes: Insights from molecular dynamics simulations. *Mater. Today Commun.* **2024**, *40*, 109591. [CrossRef]
47. Pereira, M.L.; De Sousa, J.M.; Brandão, W.H.S.; Galvão, D.S.; Fonseca, A.F.; Ribeiro, L.A. Exploring the elastic properties and fracture patterns of Me-graphene monolayers and nanotubes through reactive molecular dynamics simulations. *Chem. Phys. Lett.* **2023**, *830*, 140812. [CrossRef]
48. Li, J.; Zhang, H.; Guo, Z.; Jiang, J.-W.; Chang, T. Thermal stability of twin graphene: A Reaxff molecular dynamics study. *Appl. Surf. Sci.* **2023**, *623*, 157038. [CrossRef]
49. Santos, E.A.J.; Lima, K.A.L.; Mendonça, F.L.L.; Silva, D.A.d.; Giozza, W.F.; Junior, L.A.R. PHOTH-graphene: A new 2D carbon allotrope with low barriers for Li-ion mobility. *Sci. Rep.* **2024**, *14*, 9526. [CrossRef]
50. Plimpton, S. Fast Parallel Algorithms for Short-Range Molecular Dynamics. *J. Comput. Phys.* **1995**, *117*, 1–19. [CrossRef]
51. Stukowski, A. Visualization and analysis of atomistic simulation data with OVITO—The Open Visualization Tool. *Model. Simul. Mater. Sci. Eng.* **2010**, *18*, 015012. [CrossRef]
52. Yedla, N.; Gupta, P.; Ng, T.Y.; Geethalakshmi, K.R. Effect of loading direction and defects on the strength and fracture behavior of biphenylene based graphene monolayer. *Mater. Chem. Phys.* **2017**, *202*, 127–135. [CrossRef]
53. Wang, X.; Che, J.; Huang, W.; Linghu, J.; Hou, Z. Mechanism of crack propagation in penta-graphene. *Vacuum* **2023**, *207*, 111582. [CrossRef]
54. Mortazavi, B.; Ahzi, S. Thermal conductivity and tensile response of defective graphene: A molecular dynamics study. *Carbon* **2013**, *63*, 460–470. [CrossRef]
55. Griffith, A.A. The Phenomena of Rupture and Flow in Solids. *Philos. Trans. R. Soc. A Math. Phys. Eng. Sci.* **1920**, *A221*, 163–198.
56. Donald, W.B.; Olga, A.S.; Judith, A.H.; Steven, J.S.; Boris, N.; Susan, B.S. A second-generation reactive empirical bond order (REBO) potential energy expression for hydrocarbons. *J. Phys. Condens. Matter* **2002**, *14*, 783. [CrossRef]
57. Li, M.; Li, M.; Zheng, F. New three-dimensional orthorhombic graphene structures and their elastic properties. *Phys. B Condens. Matter* **2024**, *678*, 415753. [CrossRef]
58. Niu, M.; Miao, L.; Xu, Z.; Zhao, Y.; Hao, W.; Li, J.; Li, J.; Sui, C.; Wang, C. Tunable tensile mechanical properties of bilayer graphene through inter-layer rotation. *Diam. Relat. Mater.* **2022**, *130*, 109522. [CrossRef]
59. Zhu, C.; Ji, J.; Zhang, Z.; Dong, S.; Wei, N.; Zhao, J. Huge stretchability and reversibility of helical graphenes using molecular dynamics simulations and simplified theoretical models. *Mech. Mater.* **2021**, *153*, 103683. [CrossRef]
60. Dewapriya, M.A.N.; Rajapakse, R.K.N.D. Molecular Dynamics Simulations and Continuum Modeling of Temperature and Strain Rate Dependent Fracture Strength of Graphene with Vacancy Defects. *J. Appl. Mech.* **2014**, *81*, 081010. [CrossRef]
61. Cao, A.; Qu, J. Atomistic simulation study of brittle failure in nanocrystalline graphene under uniaxial tension. *Appl. Phys. Lett.* **2013**, *102*, 071902. [CrossRef]
62. Zhang, B.; Mei, L.; Xiao, H. Nanofracture in graphene under complex mechanical stresses. *Appl. Phys. Lett.* **2012**, *101*, 121915. [CrossRef]
63. Dewapriya, M.A.N.; Meguid, S.A.; Rajapakse, R.K.N.D. Atomistic modelling of crack-inclusion interaction in graphene. *Eng. Fract. Mech.* **2018**, *195*, 92–103. [CrossRef]
64. Zhang, T.; Li, X.; Gao, H. Designing graphene structures with controlled distributions of topological defects: A case study of toughness enhancement in graphene ruga. *Extrem. Mech. Lett.* **2014**, *1*, 3–8. [CrossRef]
65. Wei, Y.; Wu, J.; Yin, H.; Shi, X.; Yang, R.; Dresselhaus, M. The nature of strength enhancement and weakening by pentagon–heptagon defects in graphene. *Nat. Mater.* **2012**, *11*, 759–763. [CrossRef] [PubMed]
66. Sunnardianto, G.K.; Bokas, G.; Hussein, A.; Walters, C.; Moultos, O.A.; Dey, P. Efficient hydrogen storage in defective graphene and its mechanical stability: A combined density functional theory and molecular dynamics simulation study. *Int. J. Hydrogen Energy* **2021**, *46*, 5485–5494. [CrossRef]
67. Li, Y.; Datta, D.; Li, Z.; Shenoy, V.B. Mechanical properties of hydrogen functionalized graphene allotropes. *Comput. Mater. Sci.* **2014**, *83*, 212–216. [CrossRef]
68. Budarapu, P.R.; Javvaji, B.; Sutrakar, V.K.; Mahapatra, D.R.; Paggi, M.; Zi, G.; Rabczuk, T. Lattice orientation and crack size effect on the mechanical properties of Graphene. *Int. J. Fract.* **2017**, *203*, 81–98. [CrossRef]
69. Rajasekaran, G.; Kumar, R.; Parashar, A. Tersoff potential with improved accuracy for simulating graphene in molecular dynamics environment. *Mater. Res. Express* **2016**, *3*, 035011. [CrossRef]
70. Zhurkov, S.N. Kinetic concept of the strength of solids. *Int. J. Fract.* **1984**, *26*, 295–307. [CrossRef]
71. Halsey, G.; White, H.J.; Eyring, H. Mechanical Properties of Textiles, I. *Text. Res. J.* **1945**, *15*, 295–311. [CrossRef]
72. Elapolu, M.S.R.; Tabarraei, A. Mechanical and Fracture Properties of Polycrystalline Graphene with Hydrogenated Grain Boundaries. *J. Phys. Chem. C* **2021**, *125*, 11147–11158. [CrossRef]

materials

MDPI

Article

The Integration of ANN and FEA and Its Application to Property Prediction of Dual-Performance Turbine Disks

Yanqing Li [1], Ziming Zhang [2], Junyi Cheng [1,3], Zhaofeng Liu [1,3], Chao Yin [1], Chao Wang [1,4,*] and Jianzheng Guo [3,*]

1 Wedge Central South Research Institute Co., Ltd., Shenzhen 518045, China; liyanqinglyq@126.com (Y.L.); cjunyi@hotmail.com (J.C.); liuzhaofeng@csu.edu.cn (Z.L.); 18573792948@163.com (C.Y.)
2 Department of Mathematics, The University of British Columbia, Vancouver, BC V6T1Z4, Canada; ziming20@student.ubc.ca
3 State Key Laboratory of Powder Metallurgy, Central South University, Changsha 410083, China
4 Shenzhen Wedge Aviation Technology Co., Ltd., Shenzhen 518000, China
* Correspondence: wangchao1234142@hotmail.com (C.W.); 218062@csu.edu.cn (J.G.)

Abstract: Regulating the microstructure of powder metallurgy (P/M) nickel-based superalloys to achieve superior mechanical properties through heat treatment is a prevalent method in turbine disk design. However, in the case of dual-performance turbine disks, the complexity and non-uniformity of the heat treatment process present substantial challenges. The prediction of yield strength is typically derived from the analysis of microstructures under various heat treatment regimes. This method is time-consuming, expensive, and the accuracy often depends on the precision of microstructural characterization. This study successfully employed a coupled method of Artificial Neural Network (ANN) and finite element analysis (FEA) to reveal the relationship between the heat treatment process and yield strength. The coupled method accurately predicted the location specified and temperature-dependent yield strength based on the heat treatment parameters such as holding temperatures and cooling rates. The root mean square error (RMSE) and mean absolute percentage deviation (MAPD) for the training set are 50.37 and 3.77, respectively, while, for the testing set, they are 50.13 and 3.71, respectively. Furthermore, an integrated model of FEA and ANN is established using a Abaqus user subroutine. The integrated model can predict the yield strength based on temperature calculation results and automatically update material properties of the FEA model during the loading process simulation. This allows for an accurate calculation of the stress–strain state of the turbine disk during actual working conditions, aiding in locating areas of stress concentration, plastic deformation, and other critical regions, and provides a novel reliable reference for the rapid design of the turbine disk.

Keywords: artificial neural network; dual microstructure heat treatment; finite element; Ni based super alloy; turbine disks

1. Introduction

Nickel-based superalloys exhibit exceptional high-temperature strength, fatigue performance, and creep resistance, making them indispensable for critical high-temperature components in aeroengines, such as turbine disks and blades [1,2]. Advanced powder turbine disks are typically formed through Hot Isostatic Pressing (HIP), Heat Extrusion (HEX), and Isothermal Forging (IF), followed by heat treatment to tailor the microstructure [3]. The optimization of heat treatment processes to achieve the desired microstructure, thus ensuring high-performance turbine disks, has become a hot research topic.

The exceptional high-temperature mechanical properties of nickel-based P/M superalloys are mainly attributed to the γ' phase precipitation strengthening [4–6]. The ordered L12 structural γ' phase is embedded in and atomically coherent with the face-centered cubic γ phase. Therefore, the mechanical properties of the P/M superalloys will be influenced by the morphology, size, volume fraction, and distribution of the γ' phase. Most characteristics of the γ' phase can be adjusted by subsolvus or supersolvus heat treatment processes to achieve an optimal configuration. In addition, grain size also significantly affects the mechanical properties of polycrystalline alloys [7]. A fine grain structure of superalloys proves advantageous in resisting deformation at low temperatures, while also hindering the development of grain facet fatigue cracks [8]. On the contrary, superalloys with coarse grain structures exhibit superior resistance to both creep and crack growth processes, especially at elevated temperatures. In the traditional manufacturing process of the turbine disk, the microstructure and mechanical property distribution across the disk after heat treatment are approximately uniform. However, during the operation conditions of the aeroengine turbine disk, the disk hub endures a combination of low temperature and high stress, while the rim is exposed to higher temperatures and lower stress. Therefore, the turbine disk should be fabricated according to the performance requirements of specific positions on the turbine disk.

The Dual Microstructure Heat Treatment (DMHT) method was developed to produce dual-performance turbine disks [9,10]. This technology utilizes specially designed heat treatment insulation and furnaces to heat treat the turbine disk. It creates a temperature gradient in the hub and rim areas of the disk, resulting in differentiated microstructures in different areas. To address the high-stress conditions in the disk hub and the low-stress conditions in the disk rim, the utilization of the DMHT method as a heat treatment approach for the disk can effectively meet the performance requirements for the turbine disk.

The DMHT requires a well-designed insulation and heat treatment parameters to meet the performance requirements for the dual-performance turbine disk. At present, many scholars have conducted studies on predicting the mechanical performance of dual-performance turbine disks [11–15]. The strengthening mechanisms, including weak and strong dislocation coupling and anti-phase boundary, have been widely used to describe the relationship between microstructure and mechanical properties. However, when investigating the relationship between microstructure and mechanical properties through alloy strengthening mechanisms, extensive characterization of the alloy is typically required. Furthermore, the accuracy of predictions heavily depends on the precision of microstructural characterization, whereas Machine Learning (ML) is an effective alternative to overcome these issues. ML is the method that focuses on creating algorithms capable of learning from, predicting, or categorizing data which has been widely employed in the field of materials science [16–19]. For instance, ML methodologies have been employed to characterize material behaviors such as deformation [20], recrystallization [18], work-hardening [21], and even the compositional design of new alloys [22]. However, predicting the performance of DMHT turbine disks using machine learning is still challenging, and there have been no previous reports on this topic based on the authors' knowledge. In this study, we successfully employed a coupling of machine learning and finite element simulations to accurately predict DMHT disk performance. We also developed an integrated model using a user-defined subroutine to combine ANN and FEA. This model allows the simulation of the turbine disk behavior under loading conditions, where the yield strength is predicted based on temperature calculations, and then the predicted yield strength is automatically updated into the yield strength of the FEA model for subsequent stress–strain analyses. This approach effectively reduces testing costs and offers novel insights for the rapid design of turbine disk in the future.

2. Fundamentals of FEA and ANN Coupled Method

This study employs a coupled approach of the FEA model and ANN model to predict the yield strength of DMHT disks and establish an integrated model of FEA and ANN for subsequent simulation of turbine disks.

Compared to traditional methods that use alloy strengthening mechanisms to predict mechanical properties, this approach bypasses the complex and cumbersome process of building "process-microstructure-property" models. Instead, it utilizes the machine learning technique to establish direct mapping relationships between heat treatment parameters and mechanical properties. By embedding this mapping relationship into the turbine disk FEA model and using temperature histories of the DMHT simulation as the input data, the predicted property distribution of turbine disks can be achieved. Moreover, in the operating condition process simulations, the yield strength attribute of the disk FEA model can be automatically updated based on the temperature distribution during the operating process and then used for subsequent stress and strain simulations. The integrated computational framework is shown in Figure 1.

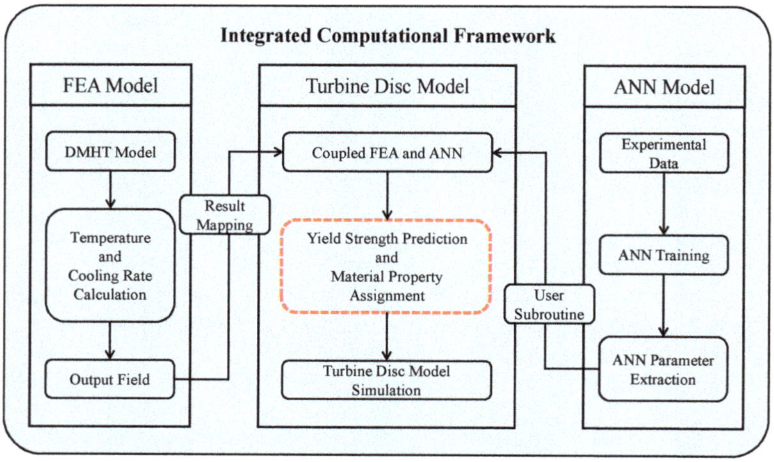

Figure 1. Integrated computational framework.

2.1. ANN Model

ANN with Back Propagation (BP) algorithms has become a vital tool in the study of supper alloy. By modeling complex, nonlinear relationships among processing conditions, microstructure, and mechanical properties, ANNs trained with BP can predict material behavior with significant accuracy. This computational approach allows researchers to circumvent traditional physical and empirical models, facilitating a more efficient design and the optimization of materials and manufacturing processes.

The input data are passed through the network from the input layer to the output layer, computing the activation at each node using the weights and activation function. If we represent activation with a and the activation function with f, the activation at each layer l can be calculated as:

$$a^{(l)} = f\left(W^{(l)}a^{(l-1)} + b^{(l)}\right) \tag{1}$$

where $W^{(l)}$ and $b^{(l)}$ are the weights and biases for the l_{th} layer, respectively.

The loss function L, such as the least absolute deviations in this study, is used to measure the difference between the network's output and the target values:

$$L = \sum_{k} |y_k - t_k| \tag{2}$$

where k is the number of values, y_k is the predicted value, and t_k is the target value.

By propagating the gradient of the loss with respect to each weight back through the network, the weights are updated. The gradient at the output layer is:

$$\delta^{(l)} = sign(y-t) \cdot f'\left(z^{(l)}\right) \tag{3}$$

where $z^{(l)}$ is the weighted sum of the l_{th} layer and $sign(x)$ is the sign function, returning +1 if $x > 0$, -1 if $x < 0$, and 0 if $x = 0$.

The gradient at the hidden layers is:

$$\delta^{(l)} = \left(W^{(l+1)T}\delta^{l+1}\right)\cdot f'\left(z^{(l)}\right) \quad (4)$$

Update weights and biases:

$$W^{(l)} = W^{(l)} - \alpha\delta^{(l)}\left(a^{(l-1)}\right)^T \quad (5)$$

where T is the matrix transpose.

$$b^{(l)} = b^{(l)} - \alpha\delta^{(l)} \quad (6)$$

Here, α is the learning rate, controlling the step size of the updates. f' represents the derivative of the activation function.

In the exploration of the heat treatment process, the relationship between the microstructure of the material and the heat treatment parameters was identified. The microstructure significantly influences the yield strength of the material. Accordingly, an ANN model was constructed with two hidden layers, utilizing heat treatment parameters as the input parameters and yield strength as the output, as illustrated in Figure 2. Given that this study involves considering a possible three-step heat treatment process, the parameters for the first step are denoted as temperature-1 and cooling rate-1, the second step parameters are labeled as temperature-2 and cooling rate-2, and the third step parameter is referred to as aging type. The tensile temperature represents the experimental temperature at which the predicted yield strength is evaluated.

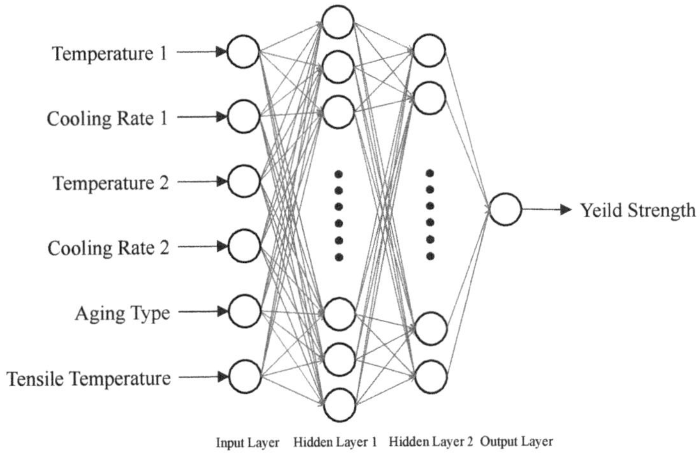

Figure 2. Architecture of an ANN model with two hidden layers.

Normalization is a pivotal step in the data processing phase. This procedure mitigates the influence of differing dimensions, enhances the precision of the model, and simplifies the underlying data structure. Prior to model training, all input and output data have been normalized to fall within the range of [0,1], a process which can be mathematically expressed as follows:

$$x' = \frac{x - x_{min}}{x_{max} - x_{min}} \quad (7)$$

As illustrated in the given expression, the normalized data are denoted by x', while x, x_{min}, and x_{max} represent the original data, the minimum value, and the maximum value of the original data, respectively. Each dimension is normalized separately within its specific range.

Adding dropout layers in an ANN model enhances its robustness and generalization performance. Dropout is a regularization technique that randomly deactivates a portion of neuron nodes during training to reduce the model's tendency to overfit. This means that, in each training iteration, only a subset of neurons participates in forward and backward propagation, making the model more capable of generalizing.

During forward propagation, for each training sample, the dropout layer randomly deactivates a fraction of neuron nodes with a probability. This can be represented as:

$$r_i \sim Bernoulli(p) \tag{8}$$

Here, r_i represents a binary random variable indicating whether the i_{th} neuron remains active, and p represents the probability of keeping a neuron active.

During backward propagation, the gradients of the deactivated neuron nodes are ignored, and only the gradients of the active neurons are considered for parameter updates.

2.2. FEA Model of DMHT

To analyze the impact of strengthening phases on material mechanical properties during the heat treatment process, it is essential to accurately measure the temperature distribution within the turbine disk during heat treatment. Traditional methods using thermocouples are unable to provide comprehensive and precise temperature measurements across the entire turbine disk. To address this challenge, this study employs finite element simulations to model the heat treatment process of the turbine disk and calculates temperature distribution and trends throughout the heat treatment process, providing valuable data for subsequent analysis [23]. The heat treatment model for the turbine disk is illustrated in Figure 3, including the turbine disk, insulation, and vacuum furnace. The DMHT process was analyzed in ABAQUS 2021. The temperature of the heat treatment and the cooling rates, both of which would be input to the ANN model, were calculated by Python scripting in the postprocess.

Figure 3. Schematic of dual-performance heat treatment components assembly.

2.3. The Integrated Model of FEA and ANN and the Analysis of Turbo Disk

To predict the yield strength of turbine disks and utilize the predicted yield strength as material properties of the FEA model for subsequent calculations, this study proposes an integrated method of FEA and ANN. The pretrained ANN model, along with the weights and biases corresponding to each neuron, was transferred into the ABAQUS 2021 user-defined subroutine USDFLD. The yield strength was linked to a user-defined field. The input data, such as HT temperatures and cooling rates, were initialized in the other user-defined fields that were mapped from the DMHT simulation results. Finally, an actual working condition analysis was performed on the turbine disk according to the aeroengine ground test conditions.

3. Materials and Experiments

In this study, a full-size dual-performance turbine disk made of the P/M superalloy FGH4113A (WZ-A3) was successfully fabricated. The nominal chemical composition is shown in Table 1. This alloy is a precipitation-strengthened nickel-based powder metallurgy superalloy, with the γ' phase serving as the principal strengthening phase.

Table 1. Chemical composition of FGH4113A alloy (wt.%).

Co	Cr	Al	Ti	W	Mo	Ta	Nb	Hf	Ni
19.0	13.0	3.0	3.7	4.0	4.0	1.0	1.2	0.2	etc.

The preparation process of the disk was vacuum argon atomization → hot isostatic pressing → hot extrusion → isothermal forging → dual-performance heat treatment. The heat treatment comprised three distinct steps. First, subsolvus heat treatment at 1120 °C was applied to ensure and improve the stability of the initial structure. Secondly, the disk underwent supersolvus heat treatment at 1200 °C with thermal insulation covering the hub to create a temperature gradient along its radial axis, promoting grain coarsening at the rim while maintaining the fine grain structure at the hub. Finally, aging heat treatment at 815 °C for 8 h improved the mechanical properties of the disk. A series of tensile tests were carried out on samples obtained from different regions of the full-size disk to ensure an accurate representation of the material's performance. Figure 4 is the sketch of the tensile specimen.

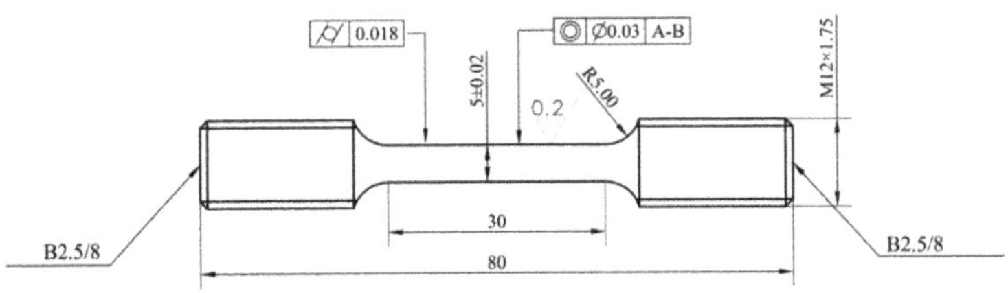

Figure 4. Sketch of tensile specimen.

The γ' phase solvus temperature is dependent on the alloy's chemical composition and can be calculated using the CALPHAD method. For the FGH4113A alloy, the solvus temperature for the γ' phase was estimated at 1152.9 °C, as shown in Figure 5. When the heat treatment temperature reached or exceeded the solvus temperature of the γ' phase, also known as supersolvus temperature, the γ' phase dissolved and reprecipitated during cooling. Controlling the cooling rate during this process allows for the adjustment of the γ' phase morphology and size distribution. Rapid quenching leads to the formation of unimodal fine γ' phases, while slow cooling results in the development of a multimodal γ' phase distribution [24–27].

The subsolvus heat treatment was conducted at a temperature below the γ' phase's solvus temperature. In this case, part of the γ' phase remains in the alloy, particularly the large primary γ' phase with a high body surface area ratio. Since the higher diffusion rate of elements at grain boundaries and the coarsening of the γ' phase following the Oswald–Ripening mechanism, there are larger γ' phases at grain boundaries. They also play a role as obstacles to limiting grain growth at relative high temperatures according to thr Zener pinning effect. However, during the subsolvus heat treatment, the majority of intergranular γ' phase dissolved. This enables the combination of fine grain structure and fine precipitates, resulting in superior performance for tensile strength and fatigue resistance up to a medium temperature [28].

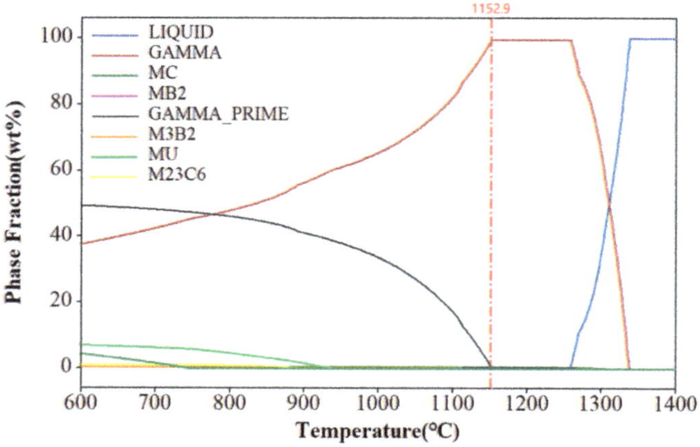

Figure 5. CALPHAD result for the FGH4113A alloy.

4. Results

4.1. Tensile Experiment Results

In this study, the ANN training dataset was constructed based on the tensile experimental results of a medium-sized experimental disk fabricated during our research process. As yield strength is primarily determined by the microstructure of the alloy after heat treatment, the heat treatment parameters were included as feature parameters in the ANN model. The thermal treatment temperature in the ANN model is defined as the temperature at the beginning of each thermal treatment cooling process, and the cooling rate is defined as the average cooling rate from the thermal treatment temperature to 800 °C. Only two aging modes were used in the heat treatment process: one involved an 815 °C hold for 8 hours, and the other involved a 760 °C hold for 16 hours. Therefore, in the ANN model, the aging types were represented by 0 and 1. These tensile specimens were subjected to a three-step heat treatment process in a vacuum furnace to simulate the heat treatment process of the full-size disk. The gas quenching pressure and fan speed were adjusted to simulate the cooling rate at different disk regions shown in Figure 6.

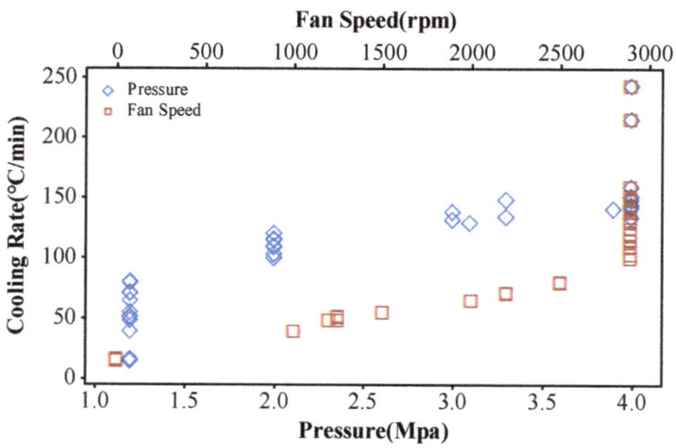

Figure 6. The relationship between cooling rate and pressure as well as fan speed in a vacuum furnace.

4.2. ANN Model Training Results

In the training of the ANN model in this study, the activation function was defined as a Rectified Linear Unit (ReLU), and the optimization was carried out using the Adaptive Moment Estimation (Adam) algorithm. Appropriate hyperparameters were carefully chosen for this ANN model, with a learning rate of 0.005 and a total of 2000 epochs.

Additionally, prior to model training, all input and output data were normalized within a [0,1] range. Dropout layers were also incorporated between the hidden layers to mitigate overfitting on the training data and enhance the model's generalization performance.

To evaluate the performance of the BP-ANN model, *RMSE* metrics and the *MADP* indicators were applied. *RMSE* and *MADP* quantify the difference between the predicted and observed values, offering insights into the model's accuracy. Lower *RMSE* and *MADP* values signify better alignment between predictions and actual observations. Together, these metrics provide a well-rounded assessment of the model's ability to predict the outcomes accurately. The formula for calculating *RMSE* and *MADP* are given by:

$$RMSE = \sqrt{\frac{1}{n}\sum_{i=1}^{n}(y_i - \hat{y}_i)^2} \quad (9)$$

$$MADP = \frac{1}{n}\sum_{i=1}^{n}\frac{(y_i - \hat{y}_i)}{\hat{y}_i} \quad (10)$$

Here, y_i is the actual value, $\hat{y}_i$ is the value predicted by the model, $\overline{y}$ is the mean of the actual values, and n represents the total amount of data.

Utilizing 150 tensile experimental sample data points from various heat treatment regimens and tensile test temperatures, a training set was constructed for the model. The relationship between partial characteristic parameters and yield strength for these samples is shown in Figure 7a–c. The dataset was divided into training and testing sets in an 8:2 ratio, following which the BP-ANN model was trained using the previously defined model parameters. The training outcomes are depicted in Figure 7d.

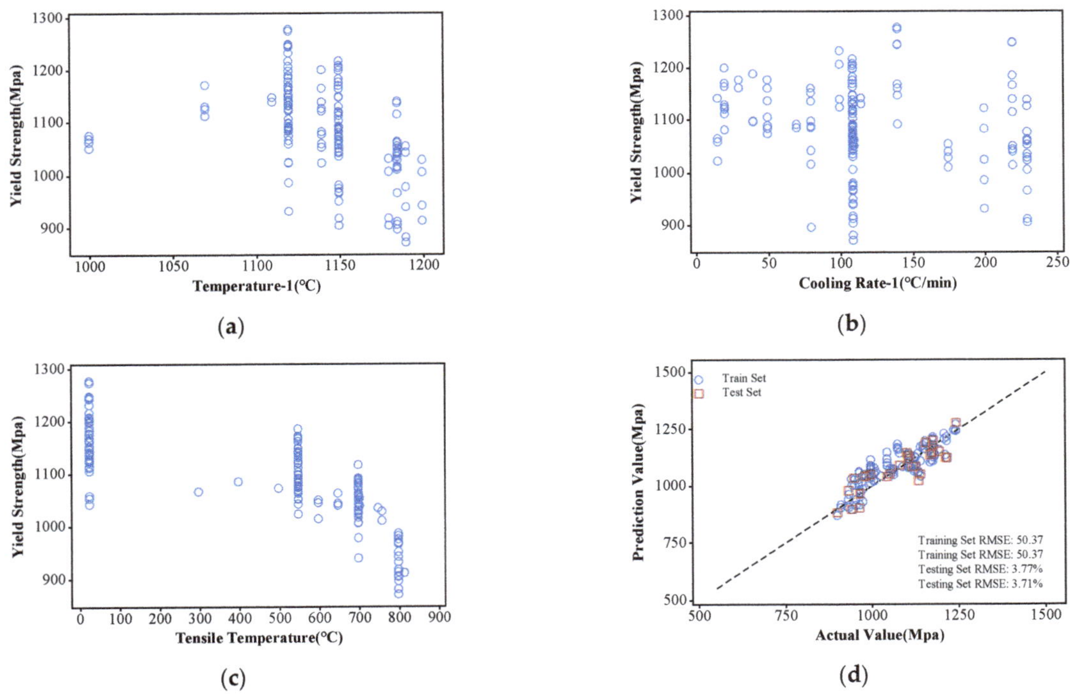

Figure 7. The partial characteristic parameters and tensile strength of the training dataset and ANN model training results: (**a**) distribution of yield strength with respect to temperature-1; (**b**) distribution of yield strength with respect to cooling rate-1; (**c**) distribution of yield strength with respect to tensile test temperature; (**d**) ANN model training results.

Model performance was assessed using *RMSE* and *MADP*, where the *RMSE* for the training set was 50.37, with an associated *MADP* of 3.77, and the testing set exhibited an *RMSE* of 50.13 with an associated *MADP* of 3.71. These results indicate a high level of accuracy and reliability relative to the model's predictions. Furthermore, the similarity in *RMSE* and *MADP* values between the training and testing sets suggests that the model shows no significant signs of overfitting and possesses strong generalization capabilities.

4.3. Finite Element Simulation Results

To ensure the accuracy of finite element simulation results, the thermal boundary conditions of the FEA model need to be calibrated before simulation. The specific method involves embedding several thermocouples in the turbine disk, referencing the measured temperature curves during the heat treatment heating and cooling process, and adjusting the model's boundary conditions to match the simulated temperature curves with the measured ones at the corresponding thermocouple positions. The placement of the thermocouples is shown in Figure 8a. A comparison of the measured temperature history curve and the simulated temperature history curve is presented in Figure 8b,c, demonstrating a good alignment between the simulated and measured temperature curves.

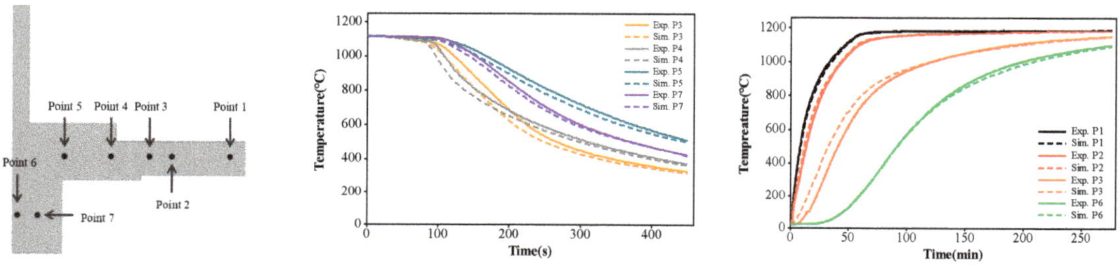

(**a**) (**b**) (**c**)

Figure 8. Finite element model calibration: (**a**) temperature measurement point locations; (**b**) temperature history curve during the heating process; (**c**) temperature history curve during the cooling process.

In accordance with the thermal treatment specifications of DMHT, finite element simulations were conducted in two distinct steps. In the first step, a uniform furnace temperature of 1120 °C was applied for the subsolvus heat treatment. Subsequently, oil quenching was performed. The holding stage temperature distribution and oil quenching cooling rate are illustrated in Figure 9a,b.

In the second step of the heat treatment, a furnace temperature of 1200 °C was applied, with the hub region covered by insulation, resulting in a radial temperature gradient across the disk. This configuration effectively subjected the rim region to a full supersolvus heat treatment while the hub region underwent a subsolvus heat treatment. Similar to step one, oil quenching was performed after supersolvus temperature holding, and the associated temperature and cooling rate are depicted in Figure 9c,d.

The simulation results indicate that, during the second stage of the heat treatment process, the transitional region attained a heat treatment temperature of approximately 1150 °C, meeting the criteria of supersolvus heat treatment for the rim region and subsolvus heat treatment for the hub region. Furthermore, both stages of heat treatment exhibited similar distribution patterns of oil quench cooling rates, characterized by lower cooling rates in the central thick region of the hub.

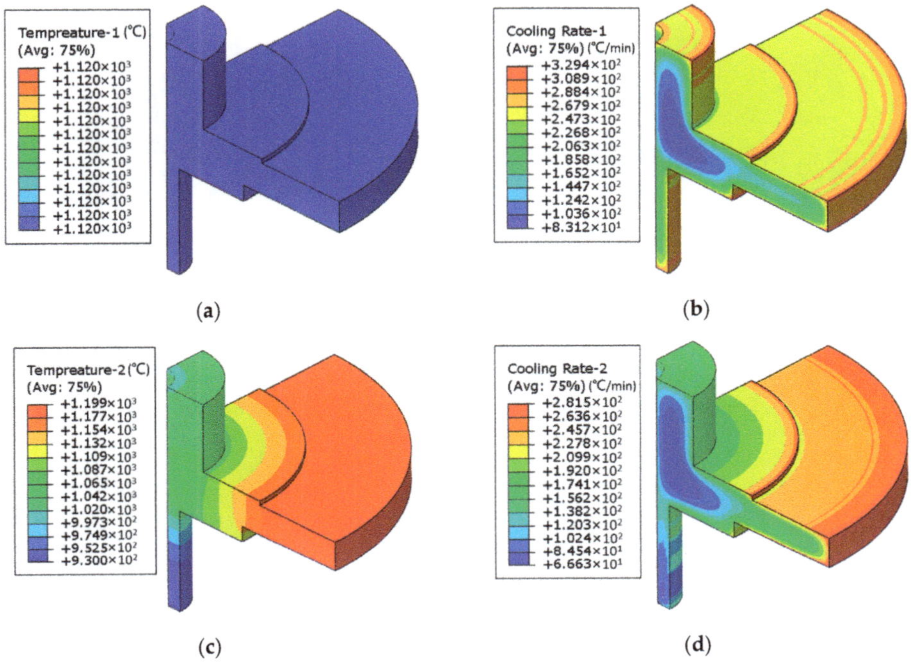

Figure 9. Finite element simulation results of DMHT: (**a**) subsolvus heat treatment temperature; (**b**) subsolvus heat treatment cooling rate; (**c**) supersolvus heat treatment temperature; (**d**) supersolvus heat treatment cooling rate.

4.4. Integration of FEA and BP-ANN

The utilization of finite element analysis to obtain temperature and cooling rate distributions within the disk during the heat treatment process, with both parameters then used as inputs for the ANN model to predict disk tensile performance, represents an efficient method to assist in the design of the heat treatment scheme. However, in the manufacturing process of turbine disks, the heat treatment process is often conducted on the blank disk, which requires further machining. As a result, the simulation results of the heat treatment process for the DMHT need to be mapped onto the final shape turbine disk model. A schematic diagram illustrating the heat-treated disk blank and the final shape of the testing disk is shown in Figure 10.

Considering the distinctive characteristics of the disk component, which exhibits varying mechanical properties across different regions, traditional finite element simulations often rely on homogenized temperature-dependent and strain-rate-dependent material properties. However, these methods fail to represent accurately the variation of material properties on the workpiece. To address this issue, this study proposes a material property assignment approach. Based on the heat treatment information for each element of FEA, material properties were calculated using the ANN model and subsequently assigned to each finite element. This coupling of ANN and FEA allowed for a precise depiction of content mechanical performance distribution across the component. The integrated computational framework of ANN and FEA is depicted in Figure 1, with the ANN model and FEA results having been established in prior research.

(**a**) (**b**)

Figure 10. Heat-treated disk blank and the final shape of the disk: (**a**) 3D model of the disk; (**b**) photo of the disk.

The integrated low-cycle fatigue calculation process described in this study follows these steps:

(1) Integration of Trained ANN model: the trained ANN model was embedded into the finite element model using a user-defined subroutine USDFLD. This included scaling coefficients, the neural network weights, and biases for each layer. The inputs and output of the ANN model are defined as user-defined field variables.
(2) Finite element results mapping: the finite element prediction results obtained from the heat-treated blank model, such as HT temperatures and cooling rates, were mapped into the final shape turbine disk model as user defined fields.
(3) Material property mapping: the yield strength was linked to a user-defined field which was predicted by the ANN model.
(4) Structural analysis: calculations were carried out based on the disk ground testing conditions. Firstly, the thermal analysis was conducted according to the tested temperature results. Secondly, a rotational load was applied to the disk for the mechanical analysis.

This integrated approach provided a comprehensive analysis of the structural behavior by considering the influence of the heat treatment process on yield strength.

Before conducting the structural analysis, it is essential to validate the model's accuracy. The model used 550 °C as the input parameter for tensile testing temperature, while the remaining input parameters were based on the mapping of the heat treatment results from the raw blank of the disk. The predicted tensile strength results of DMHT disk and the final shape of the turbine disk are shown in Figure 11.

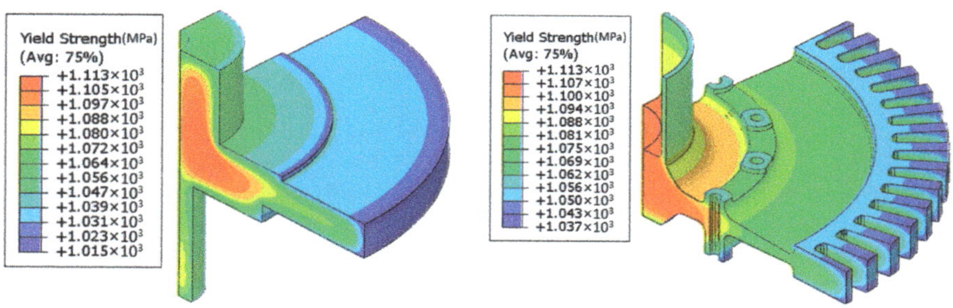

(**a**) (**b**)

Figure 11. Tensile strength prediction at 550 °C: (**a**) yield strength of DMHT disk; (**b**) yield strength of the final shape of the turbine disk.

Tensile experiments were conducted at 550 °C on specimens taken from both the hub and rim regions of the final shape turbine disk, and the results were compared with the predicted data. As shown in Figure 12, at the test temperature of 550 °C, the predicted tensile strength in the hub region was 1097 MPa, while the median experimental result was 1121 MPa, indicating a slight underprediction of 24 MPa, an error of 2.1%. In the rim region, the predicted tensile strength was 1074 MPa, compared to the experimental median of 1091 MPa, resulting in a slight underprediction of 17 MPa, with an error of 1.6%. These results demonstrate that the overall trends in the experimental and predicted results were consistent. The errors in the predictions were rather small, indicating that the model's performance was quite accurate and reliable for predicting the tensile strength.

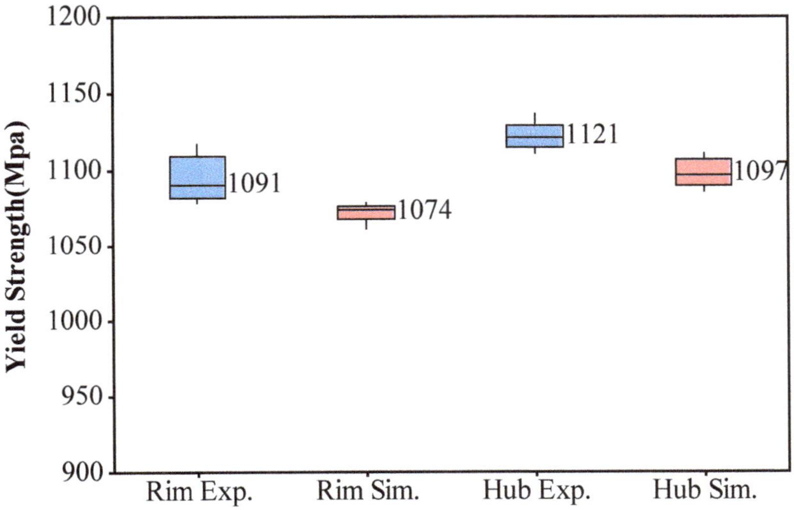

Figure 12. Yield strength prediction validation at 550 °C.

The integrated simulation of FEA and ANN was conducted under low-cycle fatigue experiment conditions of the turbine disk. The turbine disk operated at an actual speed of 15,000 rad/min, with measured temperature distributions of 320 °C at the disk hub and 570 °C at the rim. Temperature distribution and corresponding predicted tensile strength are displayed in Figure 13a,b. The finite element results calculated using the integrated model can effectively capture the uneven yield strength caused by the non-uniform temperature distribution after loading, enhancing the accuracy of subsequent stress calculations. The calculated Von Mises stress results are presented in Figure 13c, and the equivalent plastic strain results are shown in Figure 13d; minor yielding can occur, possibly, at the bolt–hole locations. This approach enabled the prediction of the yield strength distribution under actual working conditions of the turbine disk and the identification of failure regions, providing an efficient means for turbine disk heat treatment design. Additionally, by using the predicted yield strength distribution and stress state under actual working conditions, it is possible to perform topological optimization of the turbine disk geometry to reduce disk weight and enhance material utilization.

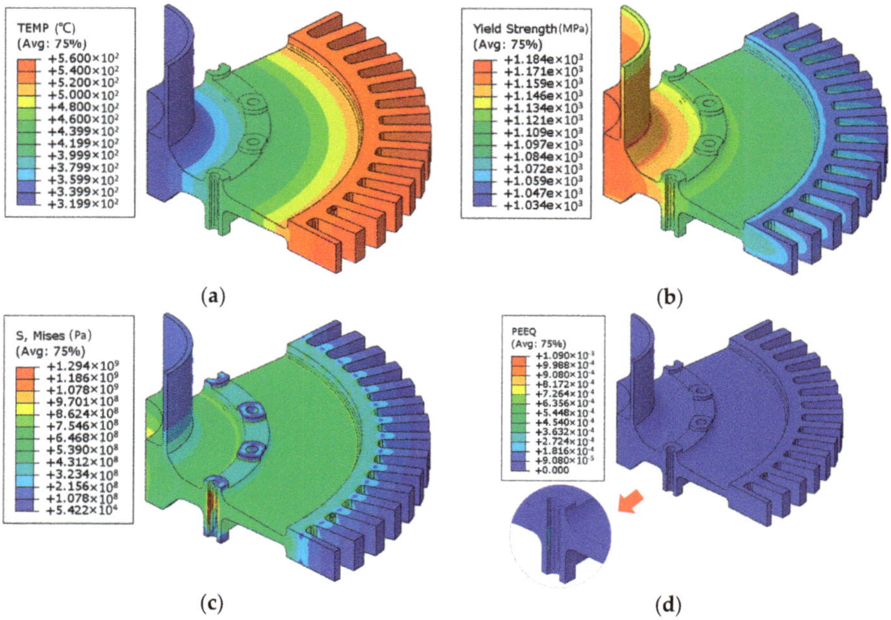

Figure 13. DMHT model for disk mapping and yield strength prediction: (**a**) temperature distribution during the test; (**b**) calculated material yield strength at the temperature of (**a**); (**c**) calculated stress distribution under testing conditions coupled with the material properties from (**b**); (**d**) calculated equivalent plastic strain distribution under testing conditions coupled with the material properties from (**b**).

5. Discussion

5.1. Qualitative Analysis of the Strengthening Mechanism

The strengthening mechanism in powder metallurgy (P/M) nickel-based superalloys primarily involves solid solution strengthening, grain boundary strengthening, and precipitation strengthening and can be summarized as follows:

$$\sigma_{ys} = \sigma_{ss} + \sigma_{gs} + \sigma_{ps} \quad (11)$$

For a specific alloy, at equilibrium state, the volume fractions of various alloy phases are temperature-dependent and remain unchanged regardless of variations in heat treatment processes. Consequently, for the F4113A alloy, the contribution of solid solution strengthening to yield strength remains constant. However, grain boundary strengthening and precipitation strengthening are highly correlated with the morphological characteristics of the alloy microstructure. Moreover, under different heat treatment regimes, the alloy microstructure exhibits significant variations, thereby becoming the primary factor influencing whether the alloy exhibits excellent tensile property or not.

To validate the relationship between microstructure and yield strength, samples were collected from different regions of the disk component for microscopic observation of their microstructures in Figure 14, including the measurement of the average grain size and average size of the secondary γ' phase.

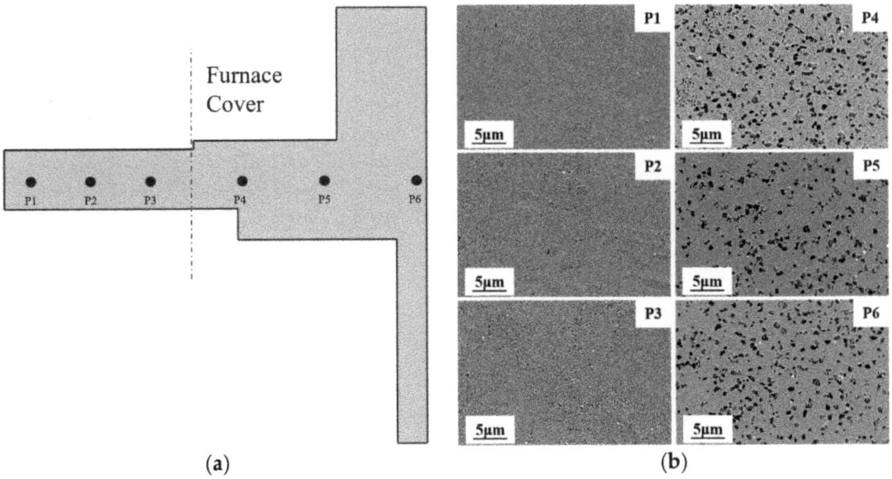

Figure 14. SEM micrographs of FGH4113A dual-performance turbine disk: (**a**) sampling point of turbine disk; (**b**) low magnification showing the γ' size in different regions of the turbine disk.

5.1.1. Solid Solution Strengthening

Solid solution strengthening in the γ and γ' phases has been commonly considered for the calculation of the overall yield strength. The chemical composition of each phase directly affects the extent of solid solution strengthening [29,30]. The contribution of solid solution strengthening to the γ and γ' phase, $\sigma_{sss}{}^{\gamma}$ and $\sigma_{sss}{}^{\gamma'}$, can be written as:

$$\sigma_{sss}{}^{\gamma} = (1 - f_{\gamma'}) \left[\sum_i (S_i^{\gamma})^2 \right]^{1/2} \tag{12}$$

$$\sigma_{sss}{}^{\gamma'} = f_{\gamma'} \sum_i \left(S_i^{\gamma'} \right)^2 \tag{13}$$

Here, $f_{\gamma'}$ is the volume fraction of the γ' phase, S_i^{γ} and $S_i^{\gamma'}$ are the degree of solid solution strengthening in the γ and γ' phases, respectively, and can be explained as:

$$S_i{}^{\gamma} = \beta_i{}^{\gamma} x_i{}^{\gamma} \tag{14}$$

$$S_i{}^{\gamma'} = \beta_i{}^{\gamma'} \left(x_i{}^{\gamma'} \right)^{1/2} \tag{15}$$

The $x_i{}^{\gamma}$ and $x_i{}^{\gamma'}$ are determined by the concentration of element i in each precipitate phase, $\beta_i{}^{\gamma}$ and $\beta_i{}^{\gamma'}$ are constants related to the atomic size and modulus.

5.1.2. Grain-Boundary Strengthening

The contribution of grain boundary strengthening to the room temperature yield strength in the FGH4113A alloy is commonly represented using the Hall–Petch relationship, expressed as follows:

$$\Delta\sigma_{gs} = f k_y d_m{}^{-1/2} \tag{16}$$

where f is the volume fraction of the γ grain, k_y represents the Hall–Petch coefficient, and d_m denotes the average grain size. According to the Hall–Petch relationship, the contribution of grain boundary strengthening to the yield strength increases as the grain size decreases. From Figure 15a, it is evident that, during the DMHT process, the insulation-covered area of the turbine disk maintained a smaller grain size. This, effectively, enhances the contribution of grain boundary strengthening to the yield strength of the turbine disk hub.

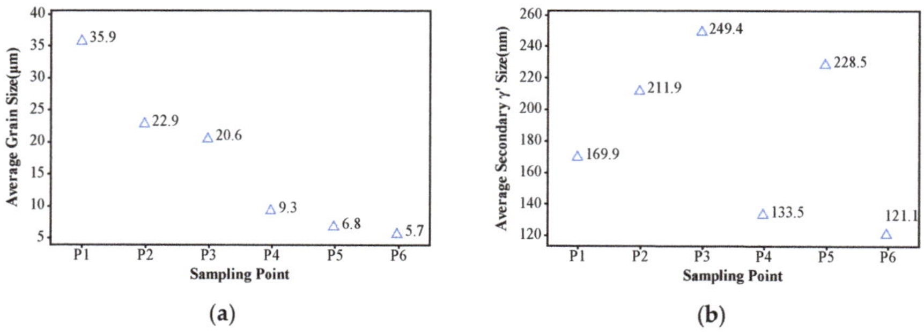

Figure 15. Statistics of the microstructure in dual-performance turbine disk: (**a**) the average grain size; (**b**) the secondary γ' size.

At high temperatures, thermal activation promotes the movement of dislocations, resulting in the activation of a greater number of slip systems compared to room temperature. Relative to the matrix, the strength of grain boundaries diminishes rapidly with increasing temperature. Under the influence of applied stress, this manifests as extensive intergranular slip, where grain boundaries lose their ability to impede dislocation motion. Concurrently, at high temperatures, grain boundaries are prone to diffusion, causing their migration, further exacerbating deformation. Consequently, grain boundary strengthening diminishes and can even vanish as temperature rises.

5.1.3. Precipitation Strengthening

The precipitation strengthening mechanisms primarily include weak pair coupling, strong pair coupling, and Orowan bowing mechanisms. The interaction between dislocations and the γ' phase is significantly influenced by the size of the γ' phase, and the enhancement of the strengthening is attributed to the formation of antiphase boundaries. When the size of the γ' phase is approximately smaller than 20 nm, multiple γ' phase particles might exist within two adjacent dislocations, a condition known as weak-pair coupling [4]. If a following dislocation enters the γ' phase while the trailing dislocation is still within the γ' phase, it is termed as strong-pair coupling [4]. The strengthening effect of the weak-pair coupling mechanism increases with a decrease in the size of the γ' phase, whereas the strong-pair mechanism exhibits the opposite behavior. As the size of the γ' phase further increases and interparticle spacing enlarges, the force required for dislocation bowing diminishes, gradually shifting the interaction mechanism with the γ' phase toward the predominant Orowan bowing mechanism.

Formulas for weak-pair coupling and strong-pair coupling are as follows [11]:

$$\Delta\tau_{weak} = \frac{1}{2}\left(\frac{\gamma_{APB}}{b}\right)^{\frac{3}{2}}\left(\frac{bd_s f}{T}\right)^{\frac{1}{2}} A - \frac{1}{2}\left(\frac{\gamma_{APB}}{b}\right) f_{\gamma'} \tag{17}$$

$$\Delta\tau_{strong} = \frac{1}{2}\left(\frac{Gb}{d_s}\right) f_{\gamma'}{}^{1/2} 0.72\omega\left(\frac{\pi d_s \gamma_{APB}}{\omega G b^2} - 1\right)^{1/2} \tag{18}$$

Among these, τ represents the critical shear stress, γ_{APB} stands for the anti-phase boundary energy per unit area, b denotes the Burgers vector, d_s signifies the mean diameter of the γ' phase, $f_{\gamma'}$ represents the volume fraction of the γ' phase, A is the geometric factor, G denotes the shear modulus of the γ' phase, ω is a parameter reflecting the elastic and it assumes a value of 1, while T is a line tensor, expressed as follows:

$$T = \frac{Gb^2}{2} \tag{19}$$

According to the formula, it is evident that the most influential variables affecting strengthening are the size and volume fraction of the γ' phase. When the γ' size meets the critical dimension, it activates both the weak-pair coupling and strong-pair coupling cutting mechanisms.

If the size of the γ' phase is large, the interaction mechanism between dislocation pairs and the γ' phase will transition from the strong-pair coupling mechanism to the Orowan bowing mechanism, satisfying the following formula:

$$\tau_{oro} = \frac{GB}{L} \tag{20}$$

where L represents the effective interparticle spacing:

$$L = \sqrt{\frac{8}{3\pi f}} d_s - d_s \tag{21}$$

At the equilibrium state, the volume fraction of the γ' phase is primarily temperature-dependent. When its size is larger, resulting in lower density, the interparticle spacing increases. Consequently, the force required for dislocation bypass around the γ' phase decreases. Calculations from the formula indicate that the critical size for weak coupling and strong coupling of the γ' phase is approximately 40 nm, while the critical size for strong coupling and Orowan bowing mechanism of the γ' phase is around 530 nm, as shown in Figure 16. When dislocations come into contact with the γ' phase, the precipitation strengthening mechanism, requiring a lower critical shear stress, is preferably initiated. Hence, when the gamma phase is below 530 nm, the precipitation strengthening mechanism is dominated by the coupling mechanism. Conversely, the Orowan bowing mechanism predominantly operates in precipitation strengthening when the gamma phase exceeds this threshold.

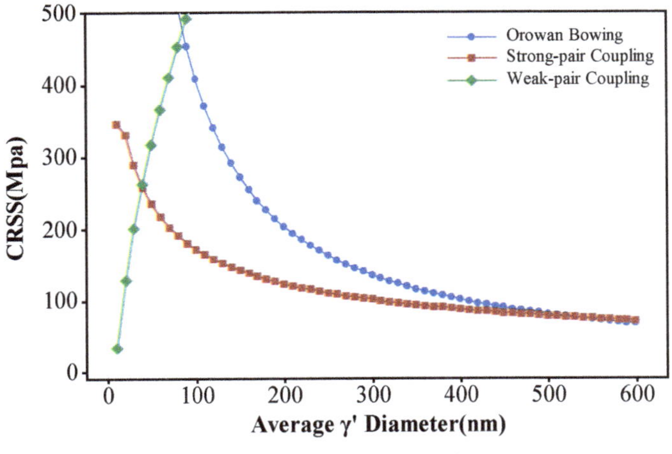

Figure 16. The critical shear stress required for the γ' phase size and the corresponding dislocation interaction mechanisms in the FGH4113A alloy.

During the DMHT process of the turbine disk, the temperature in the hub region remains lower than the alloy's solidus temperature, resulting in incomplete dissolution of the primary γ' phases into the γ matrix. As depicted in Figure 14b, numerous remnants of the primary γ' phases persist in the hub region. In contrast, the rim region exhibits the opposite behavior, as the heat treatment temperature exceeds the alloy's solidus temperature, resulting in the complete dissolution of γ' into the γ matrix.

Combining the analysis of the cooling rates, in the rim region, the heat treatment cooling rate gradually decreases radially from the center toward the edge. Consequently, the size of the precipitated γ' phases increases progressively during the cooling process. In the hub region, due to the central hole penetrating through the disk, the cooling rate in the central part, in the radial direction, is the slowest, while both the outer and inner regions exhibit faster cooling rates. This results in the largest size of the γ' phases in the central part, while the corresponding sizes in the outer and inner regions are smaller than those in the central part. Simultaneously, although the cooling rate in the hub region is slower than that in the edge region, the hub's heat treatment temperature is actually sub-solidus, allowing the undissolved primary γ' phases to occupy the elements necessary for the formation of the γ' phases. Consequently, the secondary γ' phases are smaller in the hub region compared to the rim.

From Figure 14b, it is evident that, in the rim region of the turbine disk, the γ' phases fully dissolved after the heat treatment and reprecipitated as smaller-sized secondary γ' phases. Meanwhile, in the hub region of the turbine disk, both the undissolved larger-sized primary γ' phases and the reprecipitated secondary γ' phases coexist. Additionally, as depicted in Figure 15b, in the rim region of the turbine disk, the precipitation strengthening mechanism is predominantly governed by the dislocation coupling mechanism, whereas, in the hub region, both dislocation cutting and Orowan bowing mechanisms act concurrently. Compared to the Orowan bowing mechanism, the dislocation coupling mechanism contributes more significantly to the yield strength, particularly in the hub region where the smaller grain size amplifies the effect of grain boundary strengthening. Hence, a combination of various strengthening mechanisms results in the difference in yield strength between the hub and rim regions.

5.2. Validity of the Integrated Model

This study establishes an ANN model to predict the yield strength of the turbine disk based on a dataset of tensile experimental data. The training results of the model can accurately reflect the trend of the yield strength under different heat treatment parameters, but there is still a certain error in the specific numerical value. There are several sources of error. First, in the process of establishing the model, the influence of the initial microstructure on tensile properties was not considered. In the process of material preparation, due to process limitations, both HIP and HEX processes will bring a certain degree of microstructure heterogeneity, which is one of the sources of error. Second, in the cooling process after the heat treatment, cooling is not a linear process and strengthening phases precipitate and grow faster at high temperatures. Using the average cooling rate as an input to the ANN model cannot accurately reflect the trend of material microstructure changes with decreasing temperature, which will also affect the final prediction results. Third, in the process of DMHT simulation, due to the complexity of actual working conditions, the simulation results and actual results cannot fully coincide, which also causes errors.

To address the aforementioned issues, during the training process of the ANN model, more factors affecting the tensile strength due to the heat treatment need to be considered as input parameters for the model, or the heat treatment parameters need to be refined and the amount of training set data need to be increased. In the FEA simulation of the heat treatment process, more temperature measurement points need to be added to the disk for accurate temperature measurement to calibrate the model. However, this also greatly increases the difficulty and complexity of data acquisition.

5.3. Application in Turbine Disk Design

If there is a sufficient amount of material data available, the integrated computational approach described in this study could be applicable to construct other material property models, such as creep and fatigue models, and can be used for the rapid design of dual-performance turbine disks. At the rim, fatigue life can be improved by coarse grain, which is obtained from the supersolvus heat treatment. At the hub, sub-solution heat treatment can keep the fine grain size, which is beneficial to the creep performance. By optimizing processes and the structural design, it is possible to make better use of materials at different locations, achieving a more even service life and, consequently, reducing component weight.

In the case of the need to optimize the heat treatment process for the DMHT disk, various plans involving the disk and insulation geometry and DMHT parameters can be designed for the integrated calculation. The feasibility of these plans is assessed through the calculation results and adjusted according to the property requirement of the turbine disk. For example, as described in this study, there is a stress concentration issue at the bolt–hole positions during loading, as shown in Figure 13d. However, the DMHT disk does not exhibit the highest yield strength at the bolt–hole positions, as indicated in Figure 11a. An analysis has revealed that the bolt–hole positions on the original DMHT disk are in a transitional zone between supersolvus and subsolvus heat treatment, while the supersolvus region has a higher yield strength. Therefore, one of the approaches to address this issue could be increasing the coverage area of the insulation to fully cover the heat treatment bolt–hole positions.

In the structural design of the turbine disk, calculations can be performed based on the given geometry of the turbine disk, the heat treatment process, and the loading conditions. Subsequently, structural optimizations of the turbine disk can be carried out according to the simulation results. In regions where stress concentrations occur, the original design can be retained or reinforced. In areas where there is no stress concentration, materials can be reduced to create a more structurally efficient turbine disk.

6. Conclusions

The yield strength prediction of the dual-performance turbine disk was developed using a method of integration of FEA and ANN. The following conclusions can be drawn from this study:

(1) An ANN model for predicting the yield strength of the turbine disks based on the DMHT parameters was constructed using tensile test samples from medium-sized experimental disks. RMSE and MAPD for the training set were 50.37 and 3.77, respectively, while, for the testing set, they were 50.13 and 3.71, respectively. The model exhibited good predictive performance.

(2) Through FEA, the DMHT process was simulated, and the temperature and cooling rate distribution during the DMHT process of the disk were obtained. Prior to the simulation, the thermal boundary conditions were calibrated, ensuring that the simulation results were highly reliable.

(3) A coupled method of FEA and ANN was established for yield strength prediction. The temperature and cooling rate data from the DMHT simulation results were used as input parameters for this coupled method, which successfully predicted the yield strength of the DMHT model.

(4) An integrated model of FEA and ANN was established using a user subroutine. In the loading process of the turbine disk, the yield strength was predicted based on temperature calculation results, and material properties were updated automatically. This approach successfully predicted the stress and strain states of the turbine disk during the actual loading process.

Author Contributions: Y.L.: conceptualization, investigation, software, data curation, investigation, formal analysis, methodology, visualization, writing—original draft. Z.Z.: software, formal analysis. C.W.: methodology, investigation, software, formal analysis, writing—review & editing, funding acquisition, supervision. J.C.: methodology, formal analysis. Z.L.: software, formal analysis. C.Y.: investigation, formal analysis. J.G.: methodology, investigation, writing—review & editing, funding acquisition, supervision, project administration. All authors have read and agreed to the published version of the manuscript.

Funding: This research received no external funding.

Institutional Review Board Statement: Not applicable.

Informed Consent Statement: Not applicable.

Data Availability Statement: The original contributions presented in the study are included in the article, further inquiries can be directed to the corresponding authors.

Conflicts of Interest: Authors Yanqing Li, Junyi Cheng, Zhaofeng Liu, Chao Yin and Chao Wang were employed by the company Wedge Central South Research Institute Co., Ltd. Author Chao Wang was employed by the company Shenzhen Wedge Aviation Technology Co., Ltd. The remaining authors declare that the research was conducted in the absence of any commercial or financial relationships that could be construed as a potential conflict of interest.

References

1. Thellaputta, G.R.; Chandra, P.S.; Rao, C.S.P. Machinability of Nickel Based Superalloys: A Review. *Mater. Today Proc.* **2017**, *4*, 3712–3721. [CrossRef]
2. Nowotnik, A.; Kubiak, K.; Sieniawski, J.; Rokicki, P.; Pędrak, P.; Mrówka-Nowotnik, G. Development of Nickel Based Superalloys for Advanced Turbine Engines. *MSF* **2014**, *783–786*, 2491–2496. [CrossRef]
3. Wang, Y.L.; Li, Y.; Zhang, H.; Guo, J.Z. Hot Deformation Induced Microstructure Evolution of a Novel As-Extruded Ni-Based P/M Superalloy. *Philos. Mag.* **2021**, *101*, 193–210. [CrossRef]
4. Kozar, R.W.; Suzuki, A.; Milligan, W.W.; Schirra, J.J.; Savage, M.F.; Pollock, T.M. Strengthening Mechanisms in Polycrystalline Multimodal Nickel-Base Superalloys. *Metall. Mater. Trans. A* **2009**, *40*, 1588–1603. [CrossRef]
5. Wu, Y.; Li, C.; Xia, X.; Liang, H.; Qi, Q.; Liu, Y. Precipitate Coarsening and Its Effects on the Hot Deformation Behavior of the Recently Developed γ'-Strengthened Superalloys. *J. Mater. Sci. Technol.* **2021**, *67*, 95–104. [CrossRef]
6. He, D.-G.; Lin, Y.C.; Jiang, X.-Y.; Yin, L.-X.; Wang, L.-H.; Wu, Q. Dissolution Mechanisms and Kinetics of δ Phase in an Aged Ni-Based Superalloy in Hot Deformation Process. *Mater. Des.* **2018**, *156*, 262–271. [CrossRef]
7. Zhang, X.; Chen, Y.; Cao, L.; Sun, Y.; Li, J.; Cheng, X.; Tian, G. Microstructures and Tensile Properties of a Grain-Size Gradient Nickel-Based Superalloy. *J. Alloys Compd.* **2023**, *960*, 170344. [CrossRef]
8. Tian, G.; Jia, C.; Liu, J.; Hu, B. Experimental and Simulation on the Grain Growth of P/M Nickel-Base Superalloy during the Heat Treatment Process. *Mater. Des.* **2009**, *30*, 433–439. [CrossRef]
9. Lemsky, J. *Assessment of NASA Dual Microstructure Heat Treatment Method for Multiple Forging Batch Heat Treatment*; NASA/CR—2004-212950; NASA: Washington, DC, USA, 2004.
10. Lemsky, J. *Assessment of NASA Dual Microstructure Heat Treatment Method Utilizing Ladish SuperCooler^{TM} Cooling Technology*; NASA/CR—2005-213574; NASA: Washington, DC, USA, 2005.
11. Collins, D.M.; Stone, H.J. A Modelling Approach to Yield Strength Optimisation in a Nickel-Base Superalloy. *Int. J. Plast.* **2014**, *54*, 96–112. [CrossRef]
12. Li, W.; Ma, J.; Kou, H.; Shao, J.; Zhang, X.; Deng, Y.; Tao, Y.; Fang, D. Modeling the Effect of Temperature on the Yield Strength of Precipitation Strengthening Ni-Base Superalloys. *Int. J. Plast.* **2019**, *116*, 143–158. [CrossRef]
13. Gabb, T.P.; Kantzos, P.T.; Telesman, J.; Gayda, J.; Sudbrack, C.K.; Palsa, B. Fatigue Resistance of the Grain Size Transition Zone in a Dual Microstructure Superalloy Disk. *Int. J. Fatigue* **2011**, *33*, 414–426. [CrossRef]
14. Jiang, R.; Wang, Y.C.; Zhang, L.C.; Chen, Y.; Zhang, H.; Wang, Z.B.; Song, Y.D. Fatigue Crack Propagation Behavior of the Grain Size Transition Zone in a Dual Microstructure Turbine Disc. *Int. J. Fatigue* **2023**, *172*, 107647. [CrossRef]
15. Wu, L.; Osada, T.; Watanabe, I.; Yokokawa, T.; Kobayashi, T.; Kawagishi, K. Strength Prediction of Ni-Base Disc Superalloys: Modified Γ' Hardening Models Applicable to Commercial Alloys. *Mater. Sci. Eng. A* **2021**, *799*, 140103. [CrossRef]
16. Cao, C. Prediction of Concrete Porosity Using Machine Learning. *Results Eng.* **2023**, *17*, 100794. [CrossRef]
17. Fatriansyah, J.F.; Suhariadi, I.; Fauziyyah, H.A.; Syukran, I.R.; Hartoyo, F.; Dhaneswara, D.; Lockman, Z.; Fauzi, A.; Rohman, M.S. Prediction and Optimization of Mechanical Properties of Ni Based and Fe–Ni Based Super Alloys via Neural Network Approach with Alloying Composition Parameter. *J. Mater. Res. Technol.* **2023**, *24*, 4168–4176. [CrossRef]
18. Zhu, Y.; Cao, Y.; He, Q.; Luo, R.; Zhang, J.; Di, H.; Huang, G.; Liu, Q.; Xiao, J. Machine Learning Neural-Network Identification for Dynamic Recrystallization Grains during Hot Deformation of Nickel-Based Superalloy. *Mater. Charact.* **2022**, *191*, 112108. [CrossRef]
19. Zhang, C.; Li, Y.; Jiang, B.; Wang, R.; Liu, Y.; Jia, L. Mechanical Properties Prediction of Composite Laminate with FEA and Machine Learning Coupled Method. *Compos. Struct.* **2022**, *299*, 116086. [CrossRef]
20. Lin, Y.C.; Nong, F.-Q.; Chen, X.-M.; Chen, D.-D.; Chen, M.-S. Microstructural Evolution and Constitutive Models to Predict Hot Deformation Behaviors of a Nickel-Based Superalloy. *Vacuum* **2017**, *137*, 104–114. [CrossRef]
21. Wen, H.; Jin, J.; Tang, X.; Wang, X.; Yang, H.; Zhang, Y.; Zhang, M.; Deng, L.; Wei, Q.; Chen, J.; et al. Machine Learning-Assisted Constitutive Modeling of a Novel Powder Metallurgy Superalloy. *Int. J. Mech. Sci.* **2023**, *260*, 108654. [CrossRef]
22. Gao, J.; Tong, Y.; Zhang, H.; Zhu, L.; Hu, Q.; Hu, J.; Zhang, S. Machine Learning Assisted Design of Ni-Based Superalloys with Excellent High-Temperature Performance. *Mater. Charact.* **2023**, *198*, 112740. [CrossRef]

23. Liu, Z.; Wang, C.; Cheng, J.; Guo, J. An Improved Grain Growth Model and Its Application in Gradient Heat Treatment of Aero-Engine Turbine Discs. *Materials* **2023**, *16*, 6584. [CrossRef] [PubMed]
24. Huang, G.; Liu, G.Q.; Feng, M.; Zhang, M.; Hu, B.; Wang, H. The Effect of Cooling Rates from Temperatures above the Γ′ Solvus on the Microstructure of a New Nickel-Based Powder Metallurgy Superalloy. *J. Alloys Compd.* **2018**, *747*, 1062–1072. [CrossRef]
25. Semiatin, S.L.; Mahaffey, D.W.; Levkulich, N.C.; Senkov, O.N.; Tiley, J.S. The Effect of Cooling Rate on High-Temperature Precipitation in a Powder-Metallurgy, Gamma/Gamma-Prime Nickel-Base Superalloy. *Metall. Mater. Trans. A* **2018**, *49*, 6265–6276. [CrossRef]
26. Wu, H.; Huang, Z.; Zhou, N.; Chen, J.; Zhou, P.; Jiang, L. A Study of Solution Cooling Rate on γ' Precipitate and Hardness of a Polycrystalline Ni-Based Superalloy Using a High-Throughput Methodology. *Mater. Sci. Eng. A* **2019**, *739*, 473–479. [CrossRef]
27. Zhu, L.; Pan, H.; Cheng, J.; Xiao, L.; Guo, J.; Ji, H. Dendrite Evolution and Quantitative Characterization of Γ′ Precipitates in a Powder Metallurgy Ni-Based Superalloy by Different Cooling Rates. *J. Alloys Compd.* **2022**, *918*, 165677. [CrossRef]
28. Yang, W.P.; Liu, G.Q.; Wu, K.; Hu, B.F. Influence of Sub-Solvus Solution Heat Treatment on Γ′ Morphological Instability in a New Ni–Cr–Co-Based Powder Metallurgy Superalloy. *J. Alloys Compd.* **2014**, *582*, 515–521. [CrossRef]
29. Goodfellow, A.J.; Galindo-Nava, E.I.; Schwalbe, C.; Stone, H.J. The Role of Composition on the Extent of Individual Strengthening Mechanisms in Polycrystalline Ni-Based Superalloys. *Mater. Des.* **2019**, *173*, 107760. [CrossRef]
30. Goodfellow, A.J.; Galindo-Nava, E.I.; Christofidou, K.A.; Jones, N.G.; Boyer, C.D.; Martin, T.L.; Bagot, P.A.J.; Hardy, M.C.; Stone, H.J. The Effect of Phase Chemistry on the Extent of Strengthening Mechanisms in Model Ni-Cr-Al-Ti-Mo Based Superalloys. *Acta Mater.* **2018**, *153*, 290–302. [CrossRef]

materials

MDPI

Article

Fatigue Life Data Fusion Method of Different Stress Ratios Based on Strain Energy Density

Changyin Wang [1], Jianyao Yao [1,*], Xu Zhang [1], Yulin Wu [1], Xuyang Liu [1], Hao Liu [1], Yiheng Wei [1] and Jianqiang Xin [2]

[1] College of Aerospace Engineering, Chongqing University, Chongqing 400044, China; 20176256@cqu.edu.cn (C.W.); 20176236@cqu.edu.cn (X.Z.); wuyulynn@foxmail.com (Y.W.); liuxuyang@cqu.edu.cn (X.L.); liuhaocqu@cqu.edu.cn (H.L.); 202331131075@stu.cqu.edu.cn (Y.W.)

[2] Institute for Aero Engine, Tsinghua University, Beijing 100084, China; xinjq@tsinghua.edu.cn

* Correspondence: yaojianyao@cqu.edu.cn

Abstract: To accurately evaluate the probabilistic characteristics of the fatigue properties of materials with small sample data under different stress ratios, a data fusion method for torsional fatigue life under different stress ratios is proposed based on the energy method. A finite element numerical modeling method is used to calculate the fatigue strain energy density during fatigue damage. Torsional fatigue tests under different stresses and stress ratios are carried out to obtain a database for research. Based on the test data, the W_t-N_f curves under a single stress ratio and different stress ratios are calculated. The reliability of the models is illustrated by the scatter band diagram. More than 85% of points are within ±2 scatter bands, indicating that the fatigue life under different stress ratios can be represented by the same W_t-N_f curve. Furthermore, P-W_t-N_f prediction models are established to consider the probability characteristics. According to the homogeneity of the W_t-N_f model under different stress ratios, we can fuse the fatigue life data under different stress ratios and different strain energy densities. This data fusion method can expand the small sample test data and reduce the dispersion of the test data between different stress ratios. Compared with the pre-fusion data, the standard deviations of the post-fusion data are reduced by a maximum of 21.5% for the smooth specimens and 38.5% for the notched specimens. And more accurate P-W_t-N_f curves can be obtained to respond to the probabilistic properties of the data.

Keywords: torsion fatigue; stress ratio; numerical simulation; strain energy density; data fusion

1. Introduction

Fatigue is a phenomenon in which a material is subjected to a certain number of cyclic loads at a stress level lower than the strength limit, and crack sprouting and expansion gradually occur, eventually leading to fracture failure [1,2]. Fatigue damage is one of the most common forms of failure in engineering practice. It is widely found in vehicle axles, gas turbines, aviation engines, and other major equipment [3–6]. According to statistics, more than 80% of engineering structure damage is caused by fatigue damage [7].

Solving fatigue damage problems in engineering structures requires fatigue test data to obtain large samples of data. It has been found that even if the fatigue test is conducted under the same conditions, the fatigue life obtained from the same material specimens has deviations, and the fatigue life has an unavoidable discrete nature [8–10]. The fatigue test has the features of a long cycle time and high cost; it is difficult to obtain large sample data that meet the requirements of traditional statistical methods. Therefore, to reduce the cost, it is essential to obtain an equal amount of large-sample data and a reliable fatigue life prediction model under small-sample conditions.

In recent years, data fusion methods for small sample data to obtain the same amount of large samples have been developed. The standard deviation principle [11] and the Bayes principle [12,13] are widely used in data fusion principles. The principle of standard deviation considers the conversion of standard deviation between different data levels, and the calculation is simple. However, the distribution law for small sample data is simplified. The Bayes principle can be combined with the prior information of the data to analyze the distribution law of the data. When assessing the fusion effect after data fusion, the S-N curve, or P-S-N curve, is mostly used for testing [14]. For a small sample size, it is difficult to determine the prior distribution reasonably, and the calculation is complicated. Researchers have proposed several methods for the fusion of small sample data and the fitting of P-S-N curves. Xie et al. [11] expanded the fatigue life data of aluminum alloys by standard deviation fusion based on the assumption of a lognormal distribution. The S-N and P-S-N curves were calculated, respectively, and the errors of slope and intercept after fusion were within 7%. Liu et al. [15] proposed an improved backward statistical inference method for fitting P-S-N curves, which can not only obtain reliable fatigue life but also fit more conservative P-S-N curves. Chen et al. [12] used Bayesian and hierarchical Bayesian models for fatigue data analysis and found that the root mean square error of the S-N curves plotted by the hierarchical Bayesian model was less than 5%. Klemenc et al. [16] used a two-parameter Weibull distribution to describe the data dispersion and achieve the goal of fast fitting fatigue P-S-N curves. Shimizu et al. [17] introduced the life distribution equation with three-parameter Weibull and lognormal distribution functions for the three parameters of stress-life indices, fatigue limits, and the basic dynamic stress rating, and proposed a method for analyzing the data using P-S-N curve representation for data analysis.

However, the existing data fusion methods mainly focus on the effect of stress level, and there are fewer studies on other working condition parameters (e.g., stress ratio, etc.) in the fatigue test process. Moreover, the existing fusion methods are mainly based on the statistical characteristics of the data and lack the support of physical meaning. So, the energy approach needs to be introduced to assess the physical significance of the parameters. The energy approach is an important method in the field of fatigue research. The process of fatigue is always accompanied by energy changes, and many fatigue behaviors are closely related to the absorption, consumption, and diffusion of energy [18]. This connection makes the description of fatigue through energy have a more clear physical meaning and is better combined with the stress distribution and different working conditions in the actual fatigue process. The development and research of the energy method provide a theoretical basis for data fusion methods based on more working condition parameters. Since Inglis [19] first proposed the study of energy theory, researchers have gradually linked strain energy methods with fatigue damage assessment [20–25], life prediction [26–30], consideration of notch factors [31–33], and consideration of stress ratio factors [34–36].

The stress ratio is a parameter that reflects the loading process in the fatigue test. However, it is difficult to obtain a large sample of data under any stress ratio using a fatigue test. It is necessary to fuse and expand the data on different stress ratios according to the small sample size. The energy method can provide ideas for data fusion under different stress ratios. In their research on the energy method, considering the effect of the stress ratio factor, Kadi and Ellyin [34] constructed a fatigue failure model based on strain energy and investigated the effect of stress ratios on the fatigue performance of fiberglass unidirectional panels. A standard dimensionless single curve was obtained for different combinations of stress ratios. Kujawski et al. [35] studied the effect of stress ratio on fatigue limit based on strain energy function, and the results showed that the effect of stress ratio on fatigue limit can be described in a unified form. Ellyin [36] found that when the fatigue process was analyzed by energy, the results were not sensitive to changes in the stress ratio. So when studying fatigue behavior based on energy, appropriate normalization can be performed for different stress ratios. This conclusion provides a theoretical basis for performing data fusion under different stress ratios.

Since the existing data fusion methods lack research on different stress ratio conditions, this paper innovatively proposes a data fusion method based on strain energy density for fatigue life under different strain energy density classes and different stress ratio conditions.

This method is based on the principle of normalizing the stress ratio by the energy method. The method provides ideas for the fusion and extension of small specimen data under different stress ratio conditions. This paper first establishes and validates the numerical modeling and calculation method for torsional fatigue strain energy density. Based on the fatigue life data, the W_t-N_f and P-W_t-N_f curves under a single stress ratio are calculated, respectively, and the normalized fitting models of different stress ratios are established. Finally, based on the verified data fusion method, fatigue data under different stress ratios are fused to expand the fatigue data and reduce the dispersion of the data. The probabilistic characteristics of the fatigue properties of materials are evaluated more accurately by the fused P-W_t-N_f curve.

2. Methodology

2.1. Strain Energy Density Calculation Method

To calculate the fatigue strain energy density of the material, fit the W_t-N_f curve, and study the relationship of the energy among different stress ratios, this section first introduces the basic principle of strain energy density and the method of finite element numerical modeling to calculate the strain energy density. Finally, it is validated by the data in the reference.

In this section, the fatigue process of the material is accompanied by elastic and plastic deformation, and the elastic strain energy densities ΔW_e and plastic strain energy densities ΔW_p can be calculated by the cyclic stress–strain response curve (the hysteresis loop) in the fatigue process [37,38], and the structure of the hysteresis loop is shown in Figure 1, and the principle of the calculation is expressed in Equations (1) and (2).

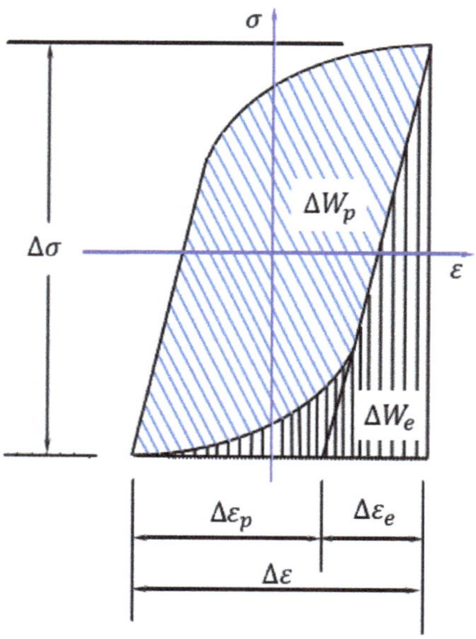

Figure 1. Schematic diagram of strain energy density [39].

$$\Delta W_e = \frac{1}{2}\Delta\sigma\Delta\varepsilon - \int_{-\frac{\Delta\sigma}{2}}^{\frac{\Delta\sigma}{2}} \varepsilon d\sigma \tag{1}$$

$$\Delta W_p = \int_{-\frac{\Delta\varepsilon}{2}}^{\frac{\Delta\varepsilon}{2}} \sigma d\varepsilon - \int_{-\frac{\Delta\sigma}{2}}^{\frac{\Delta\sigma}{2}} \varepsilon d\sigma \tag{2}$$

where ε and σ represent strain and stress, respectively, and $\Delta\varepsilon$ and $\Delta\sigma$ represent strain and stress ranges. Based on the Ramberg-Osgood formula [40], the stress–strain relationship under cyclic fatigue load can be expressed by Equation (3):

$$\varepsilon = \varepsilon_e + \varepsilon_p = \frac{\sigma}{E} + \left(\frac{\sigma}{K'}\right)^{1/n'} \tag{3}$$

where ε_p and ε_e represent plastic strain and elastic strain, E represents elastic modulus, n' represents cyclic strain hardening index, and K' is cyclic strength coefficient. For materials with Masing characteristics [41], Equations (1) and (2) can be rewritten as Equations (4) and (5):

$$\Delta W_p = \frac{1-n'}{1+n'}\Delta\sigma\Delta\varepsilon_p \tag{4}$$

$$\Delta W_e = \frac{1}{2}(\Delta\sigma\Delta\varepsilon - \frac{1-n'}{1+n'}\Delta\sigma\Delta\varepsilon_p) \tag{5}$$

The elastic strain energy density represented by Equation (4) and the plastic strain energy density represented by Equation (5) are added to obtain the total strain energy density W_t, as follows:

$$W_t = \Delta W_e + \Delta W_p = \frac{\Delta\sigma\Delta\varepsilon_p}{1+n'} + \frac{\Delta\sigma\Delta\varepsilon_e}{2} \tag{6}$$

In this paper, numerical simulation modeling is carried out using the finite element software ABAQUS 2019 to calculate the fatigue strain energy density during the fatigue process. The specific modeling and calculation flow are shown in Figure 2:

(1) Finite element modeling based on test specimens;
(2) Inputting the elastic and plastic parameters of the material;
(3) Dividing the model mesh and verifying mesh independence;
(4) Selecting the number of fatigue cycles and calculating the sinusoidal cycle curve based on the number of cycles and stress ratio;
(5) Applying the sinusoidal cyclic load to the specimen and obtaining the calculation results.

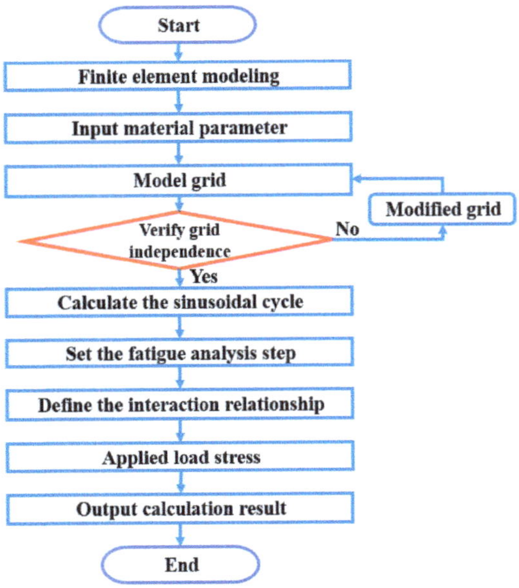

Figure 2. Strain Energy Density Modeling Calculation Flow.

The correctness of the numerical modeling approach is verified based on the data in ref. [39]. Based on the numerical modeling procedure in Figure 2, the same tensile fatigue specimens are modeled and calculated. The material constants are shown in Table 1. The simulation results are shown in Table 2. The relative error between the calculated results of this paper and those of the ref. [39] is within ±2.1%. It indicates that the method is correct.

Table 1. TC4 alloy material constants [39].

T (°C)	E (GPa)	v	n'
20	109	0.34	0.07

Table 2. Comparison between the calculated and reported strain energy densities [39].

TC4 Alloy, T = 20 °C, Kt = 1, R = −1				
Maximum Stress σ_{max}/MPa	**Fatigue Life N_f/Cycles**	**Reference Strain Energy Density W_t/(MJ/m^3)**	**Calculated Strain Energy Density W_t/(MJ/m^3)**	**Relative Error**
780	4443	12	11.7581	−2.02%
730	6262	10.3	10.32782	0.27%
680	8725	8.85	8.79789	−0.59%
630	12,441	7.63	7.56113	−0.90%
580	18,028	6.46	6.49351	0.52%
520	28,447	5.21	5.10411	−2.03%

2.2. Torsional Fatigue Test Method and Data

To investigate the relationship between strain energy density and torsional fatigue life of a certain type of alloy, torsional fatigue tests are conducted in this section to obtain the data research basis.

The tests were conducted using the torsional stress fatigue criterion at room temperature (25 ± 5 °C) in air. Smooth (Kt = 1) and notched (Kt = 3) specimens are shown in Figure 3. The specimens were manufactured based on the standards of the Aeronautical Materials Handbook [42]. The specimen was clamped in the fatigue testing machine, as shown in Figure 4a. The lower end of the tester held the specimen in place, and the upper end applied a sinusoidal load with a loading frequency of 10 Hz. The load was applied according to the maximum torsional stress. The maximum stress was adjusted by the grouping method. Stress ratios were selected as −1 and 0.05. With the stress ratio being −1, for example, the torque loading waveform is shown in Figure 4b. Torsional fatigue tests under different stresses and stress ratios were conducted according to the above test conditions. The fracture morphology under smooth and notched specimens after torsional fatigue testing is shown in Figures 5 and 6. The results show that the fracture surface of the smooth specimen has obvious crack initiation and extension zones, and the fracture surface of the notched specimen is relatively rough.

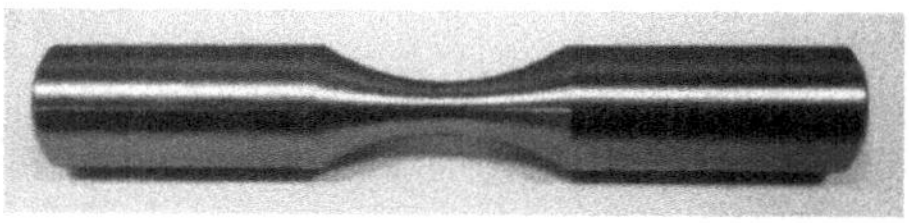

(**a**) Kt = 1

(**b**) Kt = 3

Figure 3. Schematic diagram of a room-temperature specimen.

(a) Specimen clamping

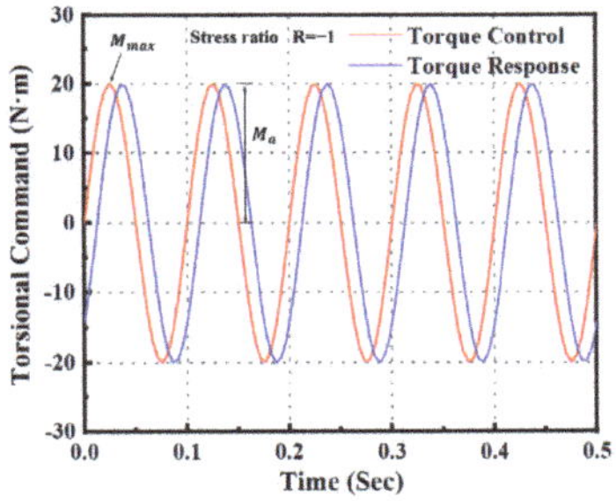

(b) loading waveforms

Figure 4. Specimen clamping and loading waveforms.

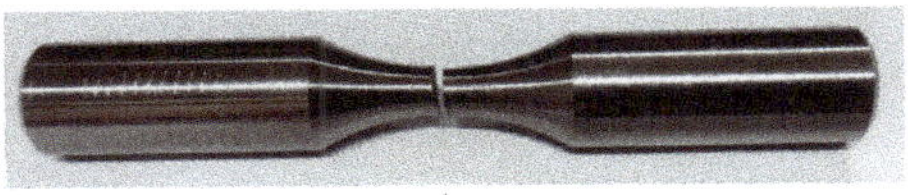

(a) Kt = 1

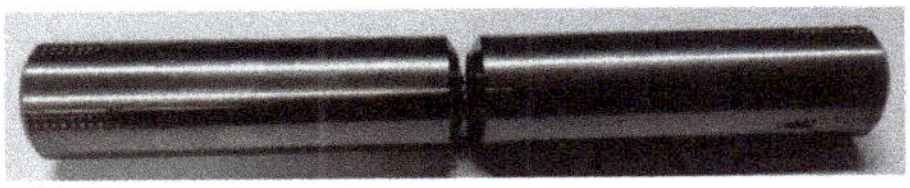

(b) Kt = 3

Figure 5. Schematic diagram of the fracture of room-temperature specimens.

(a) Kt = 1

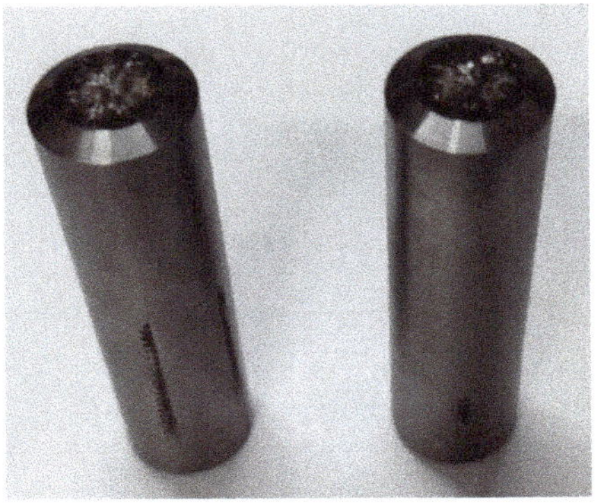

(b) Kt = 3

Figure 6. Fracture morphology of room-temperature specimens.

Based on the experimental standard working conditions and loading conditions, this section calculates the fatigue strain energy density of each working condition in torsional fatigue. Finite element models were established based on the reference standard specimen size, and the model meshes were divided, as shown in Figures 7 and 8. Table 3 shows the input of the material constant of the finite element model. When the mesh size of the finite element model is 0.8 mm, the maximum torsional stress in the S13 direction of the specimen is 535.5 MPa, as shown in Figure 9. The finite element models with different mesh sizes are calculated and compared, as shown in Figure 10. It is found that the model has converged at a mesh size of 0.8 mm, so the mesh size chosen for modeling in this paper is 0.8 mm. In setting up the interaction, the ends of the specimen are dissected and coupled to reference points to simulate the clamping of the test equipment. When applying the load, one end of the specimen is completely fixed, and the other end is loaded with a sinusoidal alternating concentrated torque load to simulate the loading of the equipment, as shown in Figures 11 and 12. The torsional fatigue data and calculated fatigue strain energy density results are shown in Tables 4 and 5.

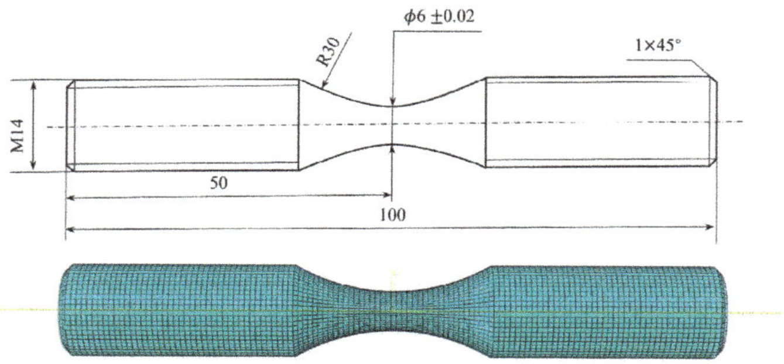

Figure 7. Smooth specimen size parameters and finite element modeling.

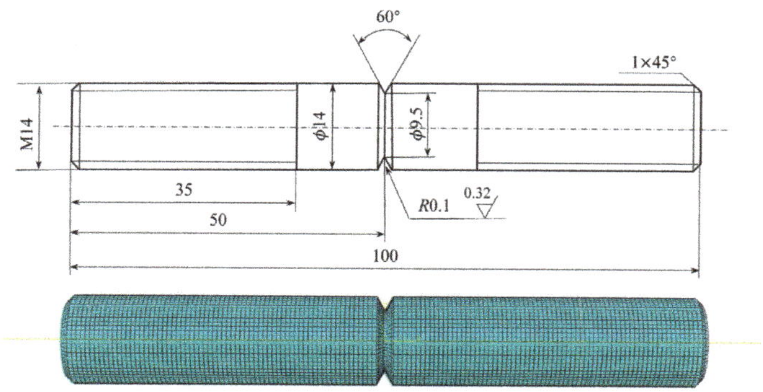

Figure 8. Notched specimen size parameters and finite element modeling.

Table 3. Material parameters for the investigated alloy [42].

T (°C)	E (GPa)	v	n'
25	196.6	0.302	0.07

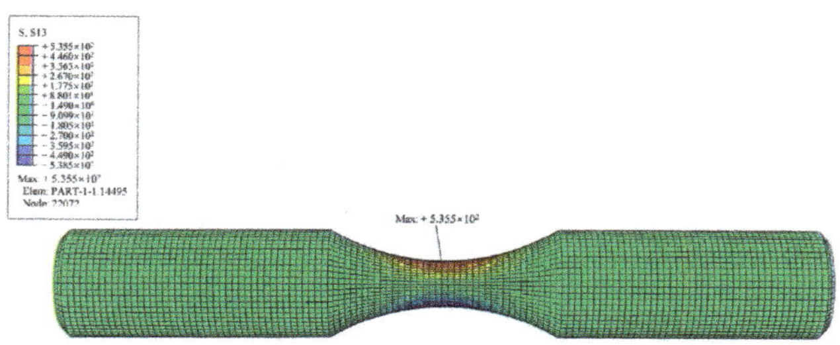

Figure 9. Finite element calculations for size 0.8 mm.

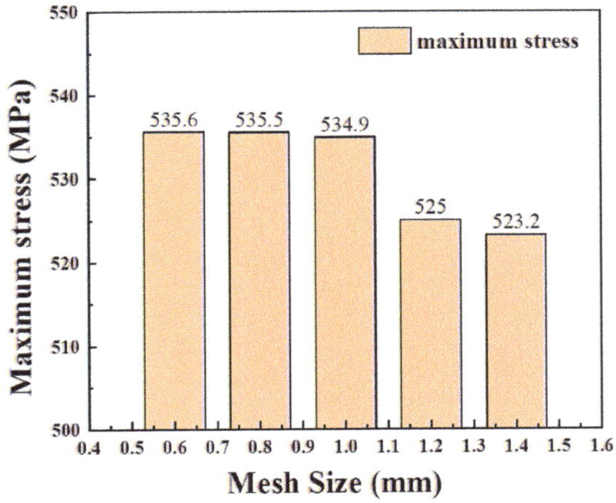

Figure 10. Finite element calculation results for different size dimensions.

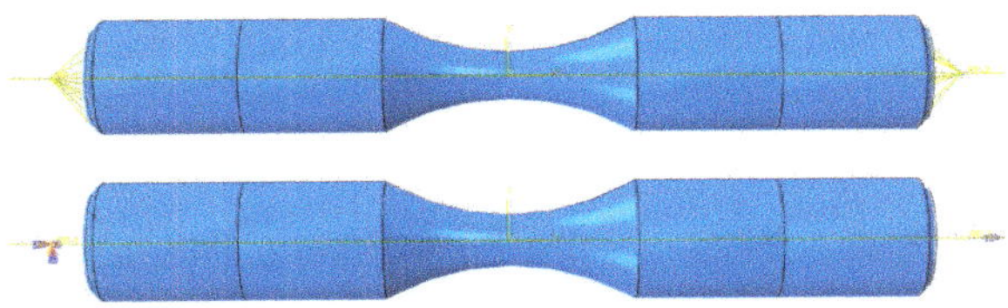

Figure 11. Smooth specimen coupling relationships and loads.

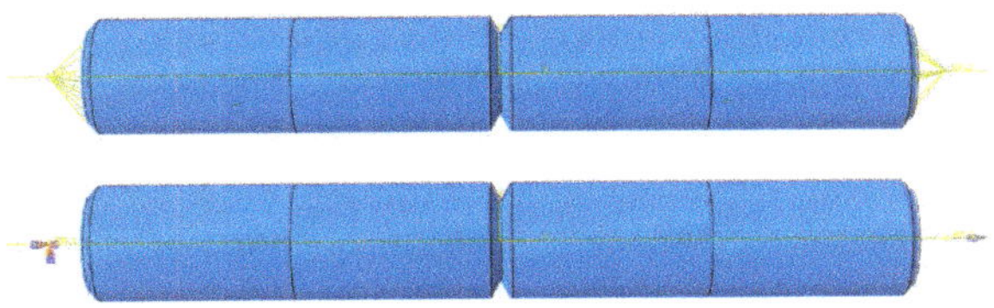

Figure 12. Notched specimen coupling relationships and loads.

Table 4. Fatigue life and strain energy density calculation results for smooth specimens.

Stress Ratio R	Maximum Stress σ_{max}/MPa	Logarithmic Median Life N_f/Cycles	Strain Energy Density W_t/(MJ/m^3)
−1	700	7447	7.2057
−1	600	39,084	4.9239
−1	550	77,983	4.0795
−1	500	232,809	3.3485
−1	400	533,335	2.0273
−1	350	803,526	1.5393
0.05	1100	15,524	23.5288
0.05	1000	26,363	14.5014
0.05	900	32,885	6.6179
0.05	850	76,560	5.5978
0.05	800	177,419	4.5450
0.05	700	276,694	3.0503
0.05	650	322,833	2.4364
0.05	600	533,188	1.9197

Table 5. Fatigue life and strain energy density calculation results for notched specimens.

Stress Ratio R	Maximum Stress σ_{max}/MPa	Logarithmic Median Life N_f/Cycles	Strain Energy Density W_t/(MJ/m^3)
−1	500	27,356	3.0272
−1	450	44,030	2.6378
−1	400	30,654	2.5113
−1	350	38,493	2.1038
−1	300	85,066	1.5463
−1	250	376,897	0.8068
−1	200	1,000,000	0.5165
0.05	700	23,287	2.3784
0.05	650	24,578	1.9174
0.05	600	50,506	1.8471
0.05	550	47,405	1.3267
0.05	500	227,879	0.9575
0.05	450	228,355	0.7662
0.05	400	968,486	0.5294

2.3. W_t- N_f Curve and P-W_t-N_f Curve Calculation Method

The strain energy density-fatigue life (W_t-N_f) curve can be solved using a two-parameter power function model [43] as shown in Equation (7) or a three-parameter power function model [44] as shown in Equation (8). Formally, the three-parameter model has one more constant term than the two-parameter model, which can express the endurance limit of the model curve. The physical meaning of the endurance limit is that the material will not undergo fatigue damage when the strain energy density is below the endurance limit. However, the calculation principle of the three-parameter model is more complicated.

$$W_t = \Delta W_e + \Delta W_p = A\left(N_f\right)^B \tag{7}$$

$$W_t = \Delta W_e + \Delta W_p = A\left(N_f\right)^B + C \tag{8}$$

The probabilistic fatigue strain energy density-fatigue life (P-W_t-N_f) curve is solved according to the one-side tolerance factor method [15]. Based on the average fatigue life x_a and standard deviation σ at each strain energy density, the probabilistic fatigue life x_p can be expressed as:

$$x_p = x_a + k_{(p,\gamma,v)} \cdot \sigma \tag{9}$$

where $k_{(p,\gamma,v)}$ is the one-side tolerance factor. p is the survival rate, which is 90% and 99% in this paper. γ is the confidence degree, which is uniformly selected as 95% in this paper. v is the degree of freedom, depending on the number of samples.

The one-side tolerance factor is related to the number of samples. When the number of samples increases, the value of $k_{(p,\gamma,v)}$ will decrease, making the standard deviation more accurate. To obtain more accurate P-W_t-N_f curves, it is necessary to expand the sample size by using reliable data fusion methods.

2.4. Data Fusion Method

In this section, the data fusion method for fatigue life of different stress ratios by energy method is presented, and the specific process is given. The fatigue life data of small samples was fused based on the point consistency principle of fatigue probability proposed in ref. [10]. The fatigue probability loci consistency principle means that no matter how the strain energy density level and stress level of the sample change, the test life of the same specimen at different levels will correspond to the same probability loci of the life distribution determined by the strain energy density level of the sample, as shown in Figure 13. The lifetime probability loci consistency principle can be expressed by Equation (10):

$$p(n_{ji}) = p(n_{ki}) \tag{10}$$

where n_{ji} represents the fatigue life of specimen i at strain energy density level j, n_{ki} represents the fatigue life of specimen i at strain energy density level k, $p(n_{ji})$ represents the probability of life being less than n_{ji} at strain energy density level j, and $p(n_{ki})$ represents the probability of life being less than n_{ki} at strain energy density level k.

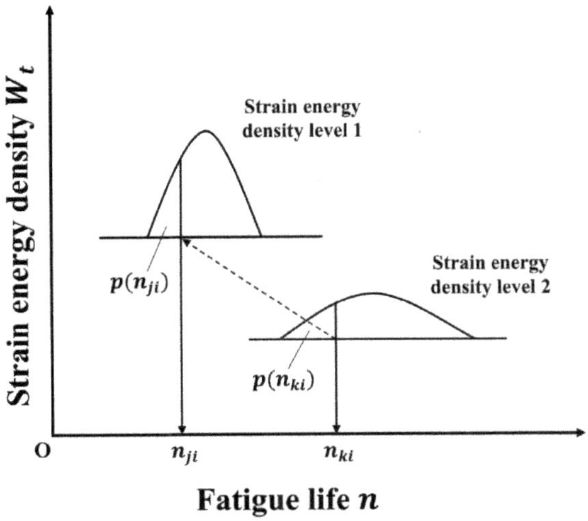

Figure 13. Lifetime probability quantile consistency schematic diagram [10].

For the test data subject to a lognormal distribution, the probability distribution is shown in Equation (11). Combined with Equation (10), Equation (12) can be obtained for mutual conversion under different strain energy densities:

$$P(N < \lg n_i) = \phi(\frac{\lg n_i - \mu_i}{\sigma_i}) \tag{11}$$

$$\frac{\lg n_{ji} - \mu_j}{\sigma_j} = \frac{\lg n_{ki} - \mu_k}{\sigma_k} \tag{12}$$

where μ_j and μ_k represent the mean logarithmic fatigue life at strain energy density levels j and k, respectively, and σ_j and σ_k represent the standard deviation of logarithmic fatigue life at strain energy density levels j and k, respectively.

According to the standard deviation fusion method [45], the relationship between the standard deviation of logarithmic fatigue life and strain energy density under equivalently large samples is shown in Equation (13):

$$\sigma_j' = \sigma_k' + K(W_{tk} - W_{tj}) \tag{13}$$

where W_{tj} and W_{tk} represent the strain energy density levels j and k, respectively, and K is the fitting coefficient, which can be used as the convergence parameter in the fusion process.

Based on the above data fusion theoretical methods, the data fusion process in this section is shown in Figure 14:

(1) Calculating the mean logarithmic fatigue life of each strain energy density level;
(2) Calculating the logarithmic life standard deviation of the fusion target level;
(3) Fusing data from other strain energy density levels and calculating the standard deviation of the fused life data;

(4) Setting an initial value of K_0 and calculating the equivalent standard deviation from other levels to the target level;
(5) Calculating the relative error Δ of two standard deviations. When Δ meets the error requirement, the fusion result is the output.

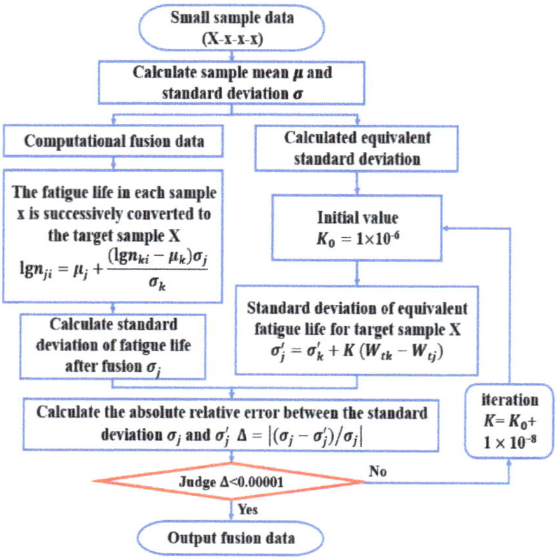

Figure 14. Data fusion flow chart.

3. Results and Discussion

3.1. W_t-N_f Curve Fitting from Test Data

For the smooth specimen shown in Figure 3, the fatigue life and strain energy density in Table 4 are fitted by two-parameter and three-parameter models, respectively. The power functions are shown in Figure 15. The expressions of the models are shown in Table 6. The results show that the life prediction error of the two models is small under different strain energy densities. Under the condition of R = 0.05, the two-parameter model has a higher R^2 value and stronger prediction ability.

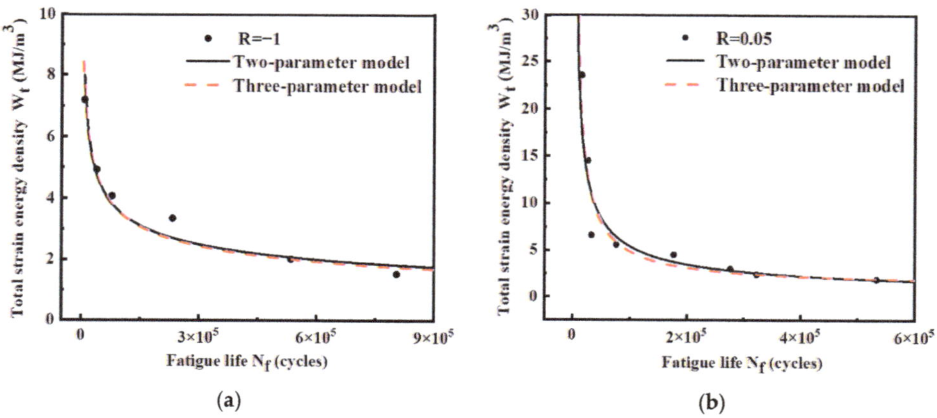

Figure 15. W_t-N_f curves of smooth specimens under different stress ratios. (**a**) Stress ratio R = −1; (**b**) stress ratio R = 0.05.

Table 6. Comparison of different model expressions and predicted life of smooth specimens.

Model Type	Different Stress Ratio Conditions	Model Expression	R^2 Value	Predicted Life of Different Strain Energy Densities/Cycles	
				5 MJ/m^3	3 MJ/m^3
Two-parameter model	R = −1	$W_t = 133.65(N_f)^{-0.315}$	0.94	33,897	171,568
	R = 0.05	$W_t = 7793.28(N_f)^{-0.631}$	0.92	114,769	257,875
	R = −1 and 0.05	$W_t = 932.65(N_f)^{-0.462}$	0.85	82,232	248,446
Three-parameter model	R = −1	$W_t = 167.07(N_f)^{-0.335}$	0.94	35,403	162,658
	R = 0.05	$W_t = 99797.05(N_f)^{-0.883} + 1.10$	0.91	98,205	221,731
	R = −1 and 0.05	$W_t = 40616.28(N_f)^{-0.825} + 1.10$	0.83	74,107	177,182

For the same specimen, the fatigue strain energy density leading to material damage is the same for the same life when the fatigue loading conditions are just different stress ratios. Based on this assumption, the strain energy density and fatigue life data under two different stress ratios are fitted using a two-parameter model and a three-parameter model, respectively. The fitted curves are shown in Figure 16. The expressions of the two models are shown in Table 6. The results show that the R^2 value of the two-parameter model is still larger than that of the three-parameter mode. Considering that the three-parameter model is more complicated and the R^2 value is smaller than that of the two-parameter model, only the two-parameter model is used below to analyze and calculate the smooth specimens.

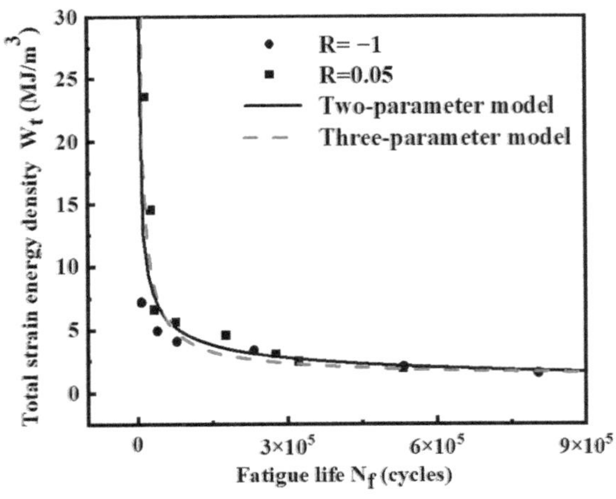

Figure 16. W_t-N_f curves of smooth specimens fitted by two stress ratios.

Based on the two-parameter model, the predicted life under different strain energy densities in the fatigue test is calculated. The scatter band diagram between the predicted life and the test life is shown in Figure 17. The results show that more than 85% of points were within ±2 scatter bands, indicating the accuracy of the two-parameter model.

For the notched specimen shown in Figure 3, according to the calculated data in Table 5, the strain energy densities are fitted with the fatigue test life through the two-parameter and three-parameter models, respectively. The power functions are shown in Figure 18. The model expressions are shown in Table 7. The results show that the life prediction error of the two models is small under different strain energy densities. The R^2 values of the two models are the same under the two stress ratios. Moreover, the fatigue life of notched specimens is smaller and more concentrated under the same strain energy density, the R^2 value of notched specimens is higher than that of smooth specimens.

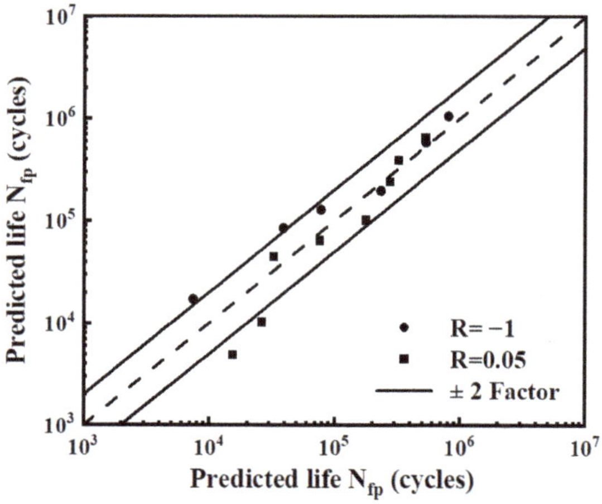

Figure 17. The scatter band diagram of the smooth specimen.

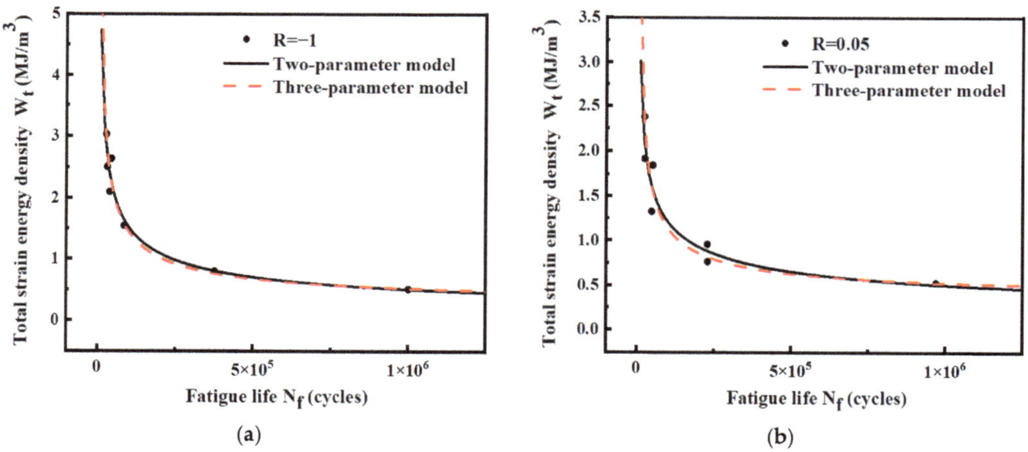

Figure 18. W_t-N_f curves of notched specimens under different stress ratios. (**a**) Stress ratio R = −1; (**b**) stress ratio R = 0.05.

The strain energy density and fatigue life data of the notched specimens under two different stress ratios are summarized. The fitting calculation is conducted by the two-parameter and three-parameter models, as shown in Figure 19. The expressions for both models are shown in Table 7. When the two stress ratios are combined, the R^2 values of the two-parameter model are higher than those of the three-parameter model. Therefore, considering the high calculation cost and low R^2 value of the three-parameter model, only the two-parameter model is used to analyze and calculate the notched specimen.

Based on the two-parameter model, the predicted life under different strain energy densities in the fatigue test is calculated. The scatter band diagram of the predicted life and the test life is shown in Figure 20. The results show that more than 85% of the points are within the ±2 scatter bands, indicating the accuracy of the two-parameter model.

Table 7. Comparison of different model expressions and predicted life of notched specimens.

Model Type	Different Stress Ratio Conditions	Model Expression	R^2 Value	Predicted Life of Different Strain Energy Densities/Cycles	
				2 MJ/m^3	1 MJ/m^3
Two-parameter model	R = −1	$W_t = 355.959(N_f)^{-0.474}$	0.97	55,926	241,379
	R = 0.05	$W_t = 95.649(N_f)^{-0.380}$	0.94	26,311	163,049
	R = −1 and 0.05	$W_t = 189.54(N_f)^{-0.429}$	0.91	40,516	203,857
Three-parameter model	R = −1	$W_t = 1472.68(N_f)^{-0.617} + 0.23$	0.97	54,050	208,289
	R = 0.05	$W_t = 942.00(N_f)^{-0.614} + 0.33$	0.94	30,269	133,967
	R = −1 and 0.05	$W_t = 4631.96(N_f)^{-0.747} + 0.37$	0.89	42,001	149,946

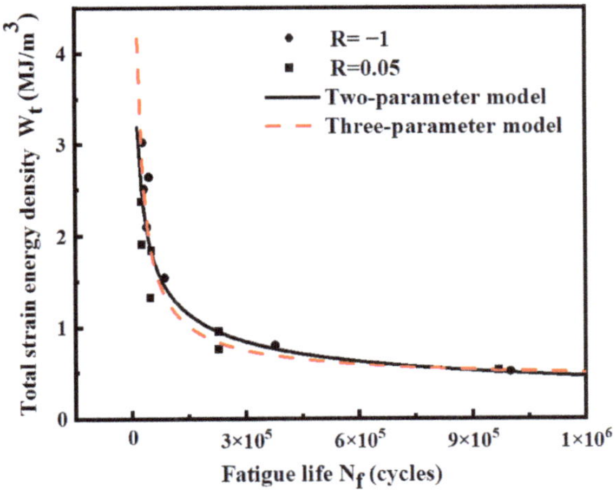

Figure 19. W_t-N_f curves of notched specimens fitted by two stress ratios.

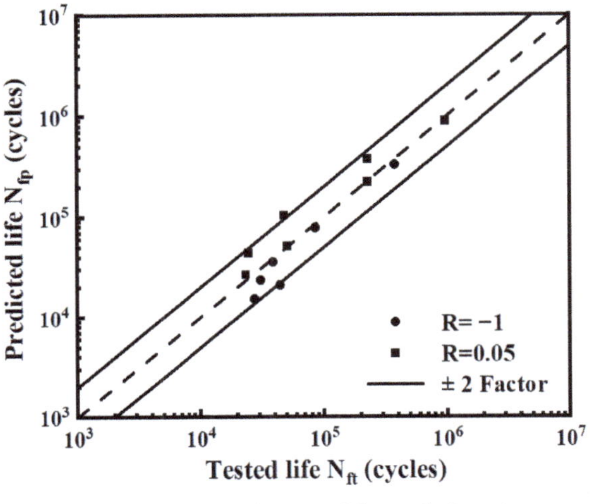

Figure 20. The scatter band diagram of the notched specimen.

3.2. P-W_t-N_f Curve Fitting

To better describe the dispersion of fatigue test data, it is of engineering significance and reference value to establish fatigue life curves with a specified probability. Using the same treatment method as P-S-N, life can be probabilized based on the obtained W_t-N_f curve.

The probability fatigue life of smooth specimens under various working conditions is calculated by Equation (9), and P-W_t-N_f curves are shown in Figures 21 and 22. The model expressions of different curves are shown in Table 8. The results show that the fatigue life predicted by the P-W_t-N_f curve is more conservative. Most of the fatigue data points are above the probability curve of P = 90%, indicating that the probability distribution of P = 90% can meet the requirements of practical engineering.

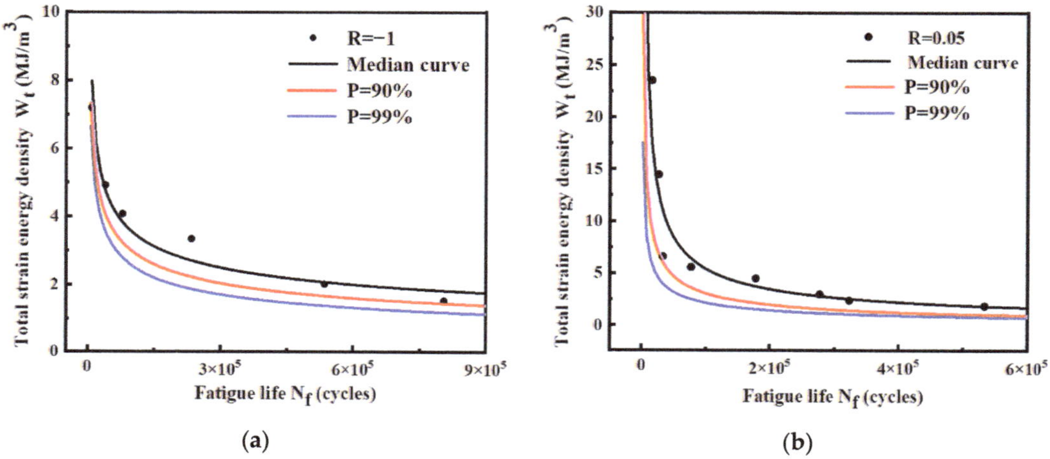

Figure 21. P-W_t-N_f curves of smooth specimens under different stress ratios. (**a**) Stress ratio R = −1; (**b**) stress ratio R = 0.05.

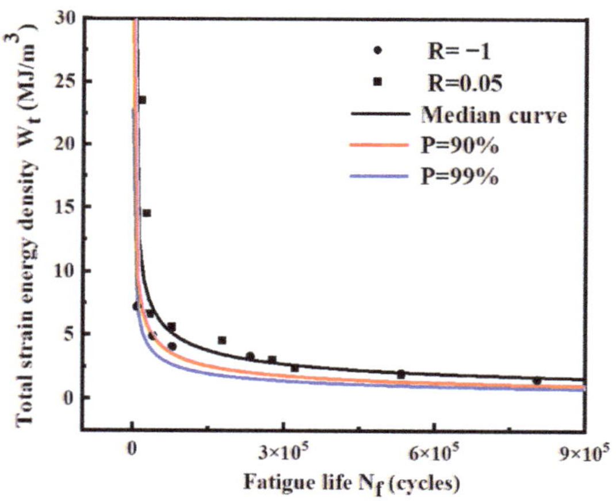

Figure 22. P-W_t-N_f curves of smooth specimens fitted by two stress ratios.

The probabilistic fatigue life of notched specimens under different conditions is calculated by Equation (9), and the P-W_t-N_f curves are shown in Figures 23 and 24. The model expressions of different curves are shown in Table 9. The results show that most fatigue data points are above the probability curve of P = 90%, indicating that the requirement can be satisfied by considering the probability distribution of P = 90% in practical engineering.

Table 8. The fitting model of the smooth specimen under different conditions.

Different Stress Ratio Conditions	Probability Level	Model Expression
R = −1	P = 90%	$W_t = 162.31(N_f)^{-0.347}$
	P = 99%	$W_t = 178.43(N_f)^{-0.368}$
R = 0.05	P = 90%	$W_t = 4026.15(N_f)^{-0.622}$
	P = 99%	$W_t = 1440.39(N_f)^{-0.563}$
R = −1 and 0.05	P = 90%	$W_t = 1008.63(N_f)^{-0.500}$
	P = 99%	$W_t = 657.25(N_f)^{-0.486}$

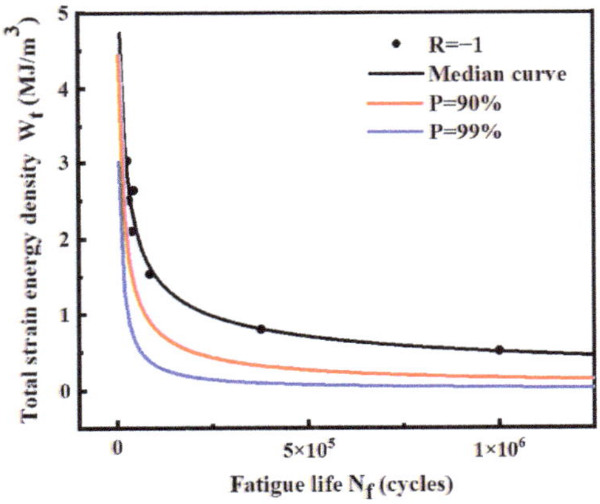

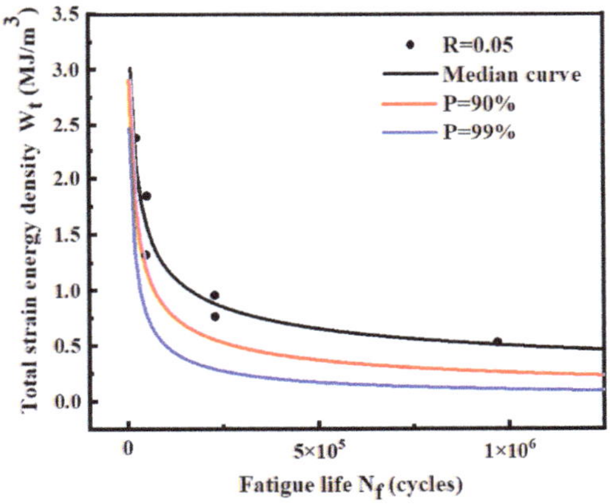

Figure 23. P-W_t-N_f curves of notched specimens under different stress ratios. (**a**) Stress ratio R = −1; (**b**) Stress ratio R = 0.05.

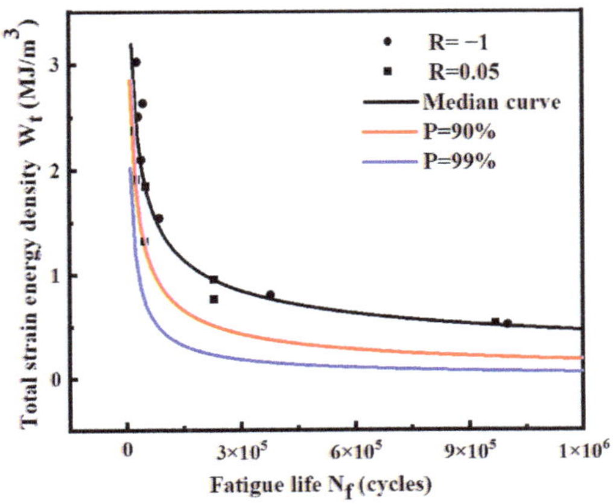

Figure 24. P-W_t-N_f curves of notched specimens fitted by two stress ratios.

Table 9. The fitting model of the notched specimen under different conditions.

Different Stress Ratio Conditions	Probability Level	Model Expression
R = −1	$P = 90\%$	$W_t = 2476.45\ (N_f)^{-0.694}$
	$P = 99\%$	$W_t = 10476.59(N_f)^{-0.894}$
R = 0.05	$P = 90\%$	$W_t = 306.27(N_f)^{-0.511}$
	$P = 99\%$	$W_t = 983.08(N_f)^{-0.657}$
R = −1 and 0.05	$P = 90\%$	$W_t = 911.55(N_f)^{-0.605}$
	$P = 99\%$	$W_t = 2775.81(N_f)^{-0.759}$

3.3. Data Fusion and P-W_t-N_f Curve Fitting Based on the Overall Data

To verify the correctness of the strain energy data fusion method, 600 fatigue test data points from ref. [46] are selected as large sample test data for calculation. According to the bending fatigue test conditions in ref. [46], three strain energy density levels are calculated. A random number program was used to randomly select five fatigue life test data points from each strain energy density level, as shown in Table 10. Table 11 shows the fatigue life data of each strain energy density level obtained using the data fusion method.

Table 10. Fatigue test data before fusion.

Strain Energy Density/(MJ/m^3)	Logarithmic Fatigue Life /Cycles
6.639	4.624, 4.699, 4.678, 4.713, 4.768
4.652	4.914, 4.992, 5.014, 5.065, 5.127
2.987	5.472, 5.532, 5.560, 5.607, 5.677

Table 11. Fatigue test data after fusion.

Strain Energy Density/(MJ/m^3)	Logarithmic Fatigue Life /Cycles
6.639	4.624, 4.699, 4.678, 4.713, 4.768, 4.593, 4.653, 4.681, 4.727, 4.798, 4.582, 4.660, 4.682, 4.734, 4.795
4.652	4.914, 4.992, 5.014, 5.065, 5.127, 4.925, 4.985, 5.013, 5.059, 5.130, 4.956, 5.001, 5.010, 5.045, 5.100
2.987	5.472, 5.532, 5.560, 5.607, 5.677, 5.461, 5.539, 5.562, 5.613, 5.674, 5.503, 5.549, 5.558, 5.592, 5.647

The mean and standard deviation of the pre-fusion data, the post-fusion data, and the overall data are shown in Table 12. Comparing the post-fusion data with the overall data, the logarithmic standard deviation of the data at each level after fusion is smaller than that of the overall data of the large sample. The maximum value of the standard deviation reduction is 30.1%. It indicates that the data fusion method can effectively expand the data volume and reduce the data dispersion.

Based on the two-parameter model, the logarithmic median W_t-N_f curve and logarithmic P-W_t-N_f curve at $P = 99\%$ of the pre-fusion data, post-fusion, and the overall data are shown in Figure 25. Only the logarithmic median W_t-N_f curve after fusion is shown in Figure 25 because the mean value of the pre-fusion data and post-fusion data is the same and the logarithmic median W_t-N_f curve coincides. The slope of the logarithmic W_t-N_f curve fitted based on the post-fusion data is −0.390, and the intercept is 2.642. The slope of the logarithmic W_t-N_f curve fitted based on the overall data is −0.382, and the intercept is 2.625. Based on the overall data, the slope relative error of the two curves is 2.08%, and the intercept relative error is 0.65%. For the probability curve of $P = 99\%$, the results show that the logarithmic P-W_t-N_f curve after fusion at a low strain energy density level is closer to the overall data than that before fusion. This is because data fusion expands the data volume and reduces the dispersion of low-strain energy density data.

Table 12. Comparison of mean and standard deviation for pre-fusion, post-fusion, and overall data.

Strain Energy Density/(MJ/m^3)	Pre-Fusion Fatigue Data N_f/Cycles		Post-Fusion Fatigue Data N_f/Cycles		Overall Large Sample Data N_f/Cycles	
	Logarithmic Mean	Logarithmic Standard Deviation	Logarithmic Mean	Logarithmic Standard Deviation	Logarithmic Mean	Logarithmic Standard Deviation
6.639	4.69046	0.053750	4.69046	0.065993	4.73836	0.067467
4.652	5.02218	0.079866	5.02218	0.065994	5.08858	0.089017
2.987	5.56984	0.077304	5.56984	0.065995	5.63849	0.094436

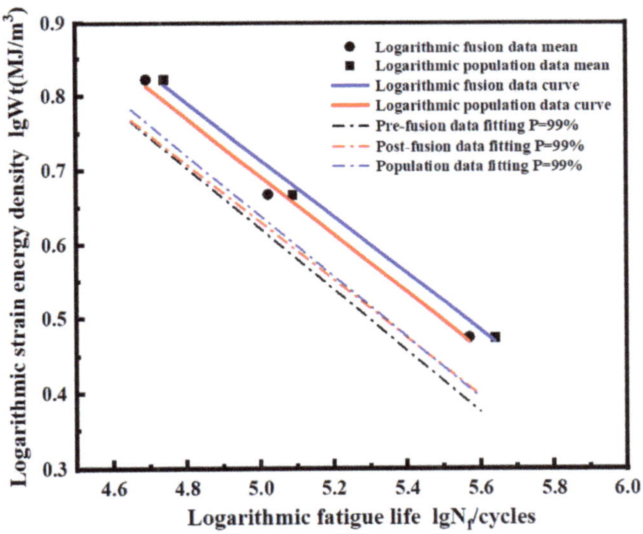

Figure 25. Comparison of logarithmic P-W_t-N_f curves of pre-fusion data, post-fusion data, and overall data.

Figure 26 shows the median W_t-N_f curve and the P-W_t-N_f curve at $P = 99\%$ of pre-fusion, post-fusion, and overall data. The expressions of each fitting curve are shown in Table 13. For the median curve, the position of the fusion curve and the overall data curve are close at the high strain energy density level and gradually deviate at the low strain energy density level. Table 13 shows the predicted life values and relative errors of the two median curves under three different strain energy density levels. Based on the overall data, the minimum relative error between the fusion data and the overall data is 11.48%, and the maximum relative error is 15.19%. The error is due to the inevitable difference between the average value of the selected small samples and the overall data. This error is within the acceptable range.

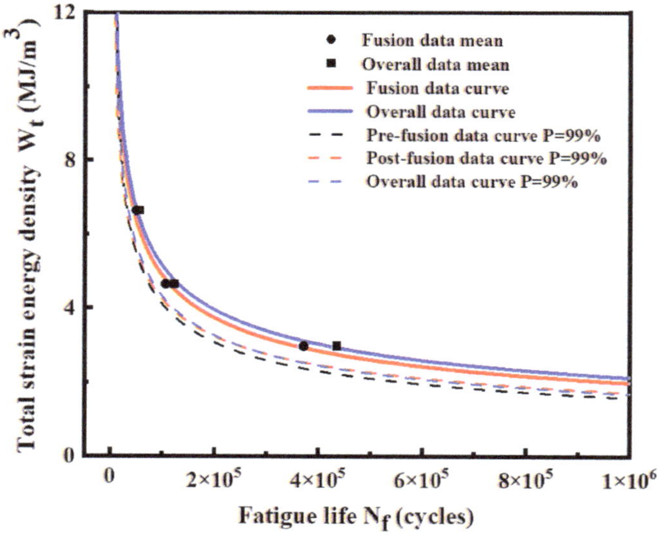

Figure 26. Comparison of P-W_t-N_f curves of pre-fusion data, post-fusion data, and overall data.

Table 13. Expressions of each probability P-W_t-N_f curve under pre-fusion, post-fusion, and overall data.

Data Type	Probability Level	Model Expression	Predicted Life of Different Strain Energy Densities and Relative Error Based on Overall Data					
			6.639 MJ/m^3	Relative Error	4.652 MJ/m^3	Relative Error	2.987 MJ/m^3	Relative Error
Pre-fusion data	$P = 50\%$	$W_t = 438.94(N_f)^{-0.390}$	46,499	11.48%	115,745	13.15%	360,451	15.19%
	$P = 99\%$	$W_t = 476.12(N_f)^{-0.411}$	32,725	6.99%	77,750	7.58%	228,473	8.31%
Post-fusion data	$P = 50\%$	$W_t = 438.94(N_f)^{-0.390}$	46,499	11.48%	115,745	13.15%	360,451	15.19%
	$P = 99\%$	$W_t = 381.36(N_f)^{-0.390}$	32,423	7.85%	80,707	4.07%	251,337	0.87%
Overall data	$P = 50\%$	$W_t = 421.98(N_f)^{-0.382}$	52,528	-	133,274	-	425,031	-
	$P = 99\%$	$W_t = 475.35(N_f)^{-0.408}$	35,184	-	84,127	-	249,181	-

For the probability curve at $P = 99\%$, results show that the P-W_t-N_f curve after fusion is closer to the P-W_t-N_f curve of the overall data than that before fusion. In Table 13, only at the high strain energy density level (6.639 MJ/m^3) is the prediction error of 7.85% after fusion larger than that of 6.99% before fusion. At the middle and low strain energy density levels (4.652 and 2.987 MJ/m^3), the prediction errors after fusion are smaller than those before fusion. Moreover, the prediction ability of the fusion data is improved continuously from a high strain energy density level to a low strain energy density level. The minimum prediction error is 0.87% at a low strain energy density level. The life prediction error between pre-fusion data and overall data is increasing. It shows that the data fusion method can expand the fatigue data at different strain energy density levels and reduce the dispersion of fatigue data at middle and low strain energy density levels. Post-fusion data are closer to the overall data, which verifies the effectiveness of the data fusion method.

3.4. Data Fusion and P-W_t-N_f Curve Fitting under Different Stress Ratios

Based on the verification of the hypothesis that stress ratio is irrelevant in strain energy density analysis of fatigue failure in Section 2.2, this section proposes to perform data fusion on logarithmic fatigue life under different stress ratios and strain energy densities. In this way, data can be expanded based on existing small sample data under different strain energy densities and stress ratios. A more accurate P-W_t-N_f curve is obtained, which provides a reference value for engineering applications.

For smooth specimens, according to the test data in Table 4, two stress ratios and two strain energy density levels for each stress ratio are selected for data fusion. The pre-fusion data and post-fusion data are shown in Table 14. The comparison of the logarithmic mean and logarithmic standard deviation before and after fusion is shown in Table 15. After the fusion of different stress ratios, the average logarithmic life of each strain energy density level remains unchanged. The standard deviation after fusion at each strain energy density level is smaller. Compared with the pre-fusion data, the maximum reduction in the standard deviation of the post-fusion data is 21.5%. The results show that the data fusion method reduces data dispersion while expanding the data.

Table 14. Fatigue life data of smooth specimens under different stress ratios before and after fusion.

Stress Ratio R	Strain Energy Density W_t/(MJ/m^3)	Logarithmic Fatigue Life before Fusion N_f/Cycles	Logarithmic Fatigue Life after Fusion N_f/Cycles
−1	4.079	4.971, 5.039, 4.664	4.971, 5.039, 4.664, 4.621, 4.971, 5.083, 4.892, 4.661, 5.122, 5.130, 4.629, 4.916
	2.027	5.456, 5.806, 5.919	5.456, 5.806, 5.919, 5.807, 5.875, 5.500, 5.727, 5.497, 5.958, 5.965, 5.465, 5.752
0.05	5.598	4.884, 4.653, 5.114	4.884, 4.653, 5.114, 4.964, 5.031, 4.656, 4.613, 4.963, 5.076, 5.122, 4.621, 4.908
	3.050	5.681, 5.180, 5.467	5.681, 5.180, 5.467, 5.522, 5.590, 5.215, 5.171, 5.521, 5.634, 5.442, 5.212, 5.673

Table 15. Comparison of mean and standard deviation of smooth specimens under different stress ratios before and after fusion.

Stress Ratio R	Strain Energy Density W_t/(MJ/m^3)	Pre-Fusion Data N_f/Cycles		Post-Fusion Data N_f/Cycles	
		Logarithmic Mean	Logarithmic Standard Deviation	Logarithmic Mean	Logarithmic Standard Deviation
−1	4.079	4.88153	0.1997664	4.88153	0.1975206
	2.027	5.71650	0.2414045	5.71650	0.1975221
0.05	5.598	4.89483	0.2306621	4.88483	0.1975220
	3.050	5.44238	0.2513805	5.44238	0.1975213

The P-W_t-N_f curves of smooth specimens before and after fusion in logarithmic coordinates are calculated, as shown in Figure 27. The results show that the fitted curve after fusion is above the curve before fusion under both probabilities P. The two curves are close at a high strain energy density level, while the deviation increases gradually at a low strain energy density level.

Figure 28 shows the distribution of data and the P-W_t-N_f curve before and after fusion. The expressions of each fitting curve are shown in Table 16. Most of the fused data are evenly distributed among the original data, which ensures that the mean fatigue life at the same strain energy density level after fusion remains unchanged, reduces the dispersion of low strain energy level data, and reduces the standard deviation.

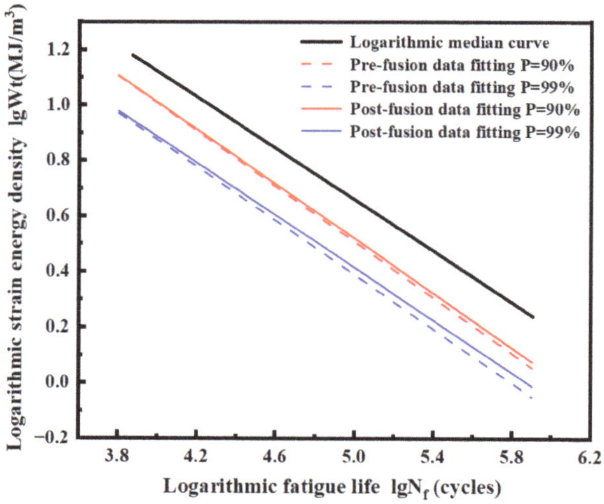

Figure 27. Median W_t-N_f curve and logarithmic P-W_t-N_f curve of smooth specimens before and after fusion.

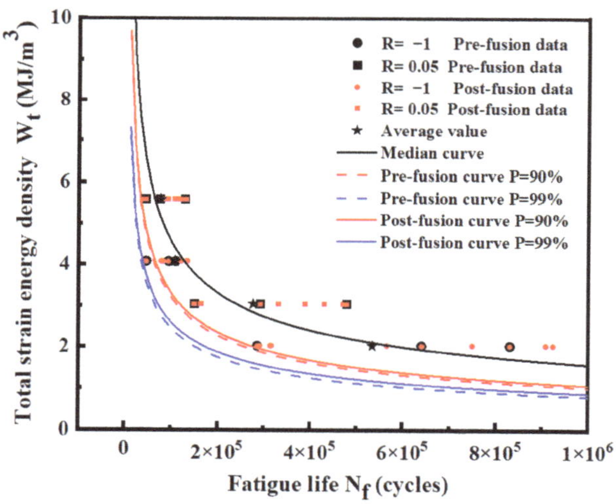

Figure 28. Data distribution and P-W_t-N_f curves of smooth specimens before and after fusion.

Table 16. Expressions of probability P-W_t-N_f curves of smooth specimens before and after fusion.

Data Type	Probability Level	Model Expression	Predicted Life of Different Strain Energy Densities N_f/Cycles			
			5.598 MJ/m^3	4.079 MJ/m^3	3.050 MJ/m^3	2.027 MJ/m^3
Pre-fusion data	$P = 90\%$	$W_t = 1008.63(N_f)^{-0.500}$	32,464	61,144	109,361	247,603
	$P = 99\%$	$W_t = 657.25(N_f)^{-0.486}$	18,140	34,795	63,285	146,694
Post-fusion data	$P = 90\%$	$W_t = 934.74(N_f)^{-0.491}$	33,636	64,091	115,860	266,275
	$P = 99\%$	$W_t = 590.47(N_f)^{-0.471}$	19,745	38,668	71,682	170,668

The predicted lives of the two curves before and after fusion under different strain energy densities are listed in Table 16. The results show that the curve after fusion gradually deviates from the curve before fusion, and the predicted life of the curve after fusion is higher from high strain energy density to low strain energy density, which is similar to the overall data in Section 3.1. After fusion, the low strain energy density level with a smaller standard deviation is closer to the real overall data, and the life prediction is more accurate. Based on the P-W_t-N_f curve given in Figure 28, the corresponding life can be obtained by calculating the strain energy density under different stress ratios.

For notched specimens, data fusion is also performed according to the test data in Table 5. The data before and after fusion are shown in Table 17, and the comparison of logarithmic mean and logarithmic standard deviation before and after fusion is shown in Table 18. The results show that after the fusion of different stress ratios, the average logarithmic life of each strain energy density level remains unchanged. In terms of standard deviation, after data fusion expands the sample data volume, the standard deviation between the data is reduced at the low strain energy density level (such as 0.807 MJ/m^3). Compared with the pre-fusion data, the maximum reduction in the standard deviation of the post-fusion data is 38.5%. At the high strain energy density stage (such as 1.917 MJ/m^3), the fatigue life of the notched specimens at fracture is small and relatively concentrated due to the large stress concentration coefficient. The standard deviation between the stages will be slightly increased after data fusion to expand the sample data. In the case of notched specimens, the data fusion method can not only expand the data but also reduce the dispersion of low-strain energy density level data.

Table 17. Fatigue life data of notched specimens under different stress ratios before and after fusion.

Stress Ratio R	Strain Energy Density W_t/(MJ/m^3)	Logarithmic Fatigue Life before Fusion N_f/Cycles	Logarithmic Fatigue Life after Fusion N_f/Cycles
−1	2.511	4.277, 4.483, 4.699	4.277, 4.483, 4.699, 4.667, 4.763, 4.029, 4.646, 4.442, 4.371, 4.115, 4.731, 4.613
	0.807	5.757, 5.853, 5.119	5.757, 5.853, 5.119, 5.367, 5.573, 5.789, 5.736, 5.531, 5.461, 5.205, 5.821, 5.703
0.05	1.917	4.550, 4.346, 4.275	4.550, 4.346, 4.275, 4.182, 4.387, 4.603, 4.571, 4.667, 3.933, 4.019, 4.635, 4.517
	0.957	4.986, 5.602, 5.484	4.986, 5.602, 5.484, 5.149, 5.354, 5.570, 5.538, 5.634, 4.901, 5.518, 5.313, 5.243

Table 18. Comparison of mean and standard deviation of notched specimens under different stress ratios before and after fusion.

Stress Ratio R	Strain Energy Density W_t/(MJ/m^3)	Pre-Fusion Data N_f/Cycles		Post-Fusion Data N_f/Cycles	
		Logarithmic Mean	Logarithmic Standard Deviation	Logarithmic Mean	Logarithmic Standard Deviation
−1	2.511	4.48648	0.2106861	4.47548	0.2451951
	0.807	5.57622	0.3987908	5.57735	0.2451965
0.05	1.917	4.39054	0.1428186	4.38061	0.2451956
	0.957	5.35770	0.3268738	5.35770	0.2451963

The P-W_t-N_f curve of the pre-fusion data and post-fusion data of the notched specimen in logarithmic coordinates is calculated, as shown in Figure 29. Under both probabilities P, the fitted curve after fusion is above the curve before fusion. The two curves are close at a high strain energy density level, and the deviation increases at a low strain energy density level.

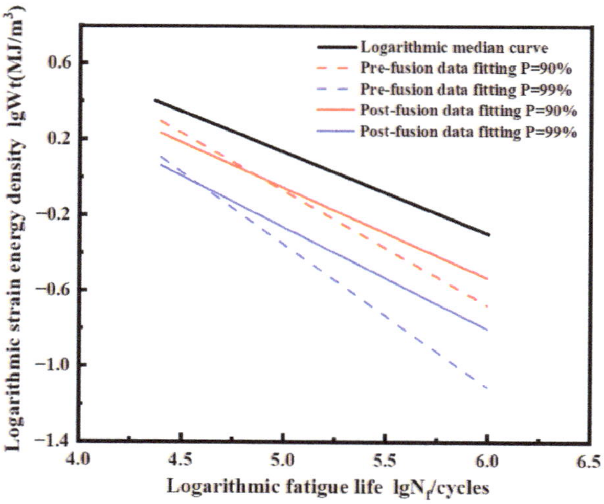

Figure 29. Median W_t-N_f curve and logarithmic P-W_t-N_f curve of notched specimens before and after fusion.

Figure 30 shows the distribution of data and the P-W_t-N_f curve before and after fusion. The expression of the probability curve P = 90% and P = 99% before and after fusion is shown in Table 19. Most of the fused data are evenly distributed among the original data. At low strain energy density levels, the expansion of sample data reduces the standard deviation of each level and decreases the dispersion of data. At a high strain energy density level, the data dispersion will slightly increase after fusion.

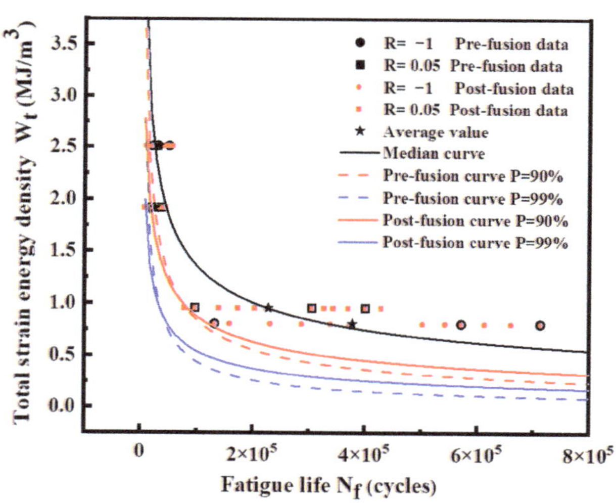

Figure 30. Data distribution and P-W_t-N_f curves of notched specimens before and after fusion.

Table 19. Expressions of probability P-W_t-N_f curves of notched specimens before and after fusion.

Data Type	Probability Level	Model Expression	Predicted Life of Different Strain Energy Densities N_f/Cycles			
			2.511 MJ/m^3	1.917 MJ/m^3	0.957 MJ/m^3	0.807 MJ/m^3
Pre-fusion data	$P = 90\%$	$W_t = 911.55(N_f)^{-0.606}$	16,762	26,167	82,343	109,094
	$P = 99\%$	$W_t = 2775.81(N_f)^{-0.759}$	10,231	14,601	36,467	45,650
Post-fusion data	$P = 90\%$	$W_t = 214.78(N_f)^{-0.477}$	11,236	19,787	84,897	121,370
	$P = 99\%$	$W_t = 271.79(N_f)^{-0.539}$	5948	9814	35,614	48,863

The predicted lives of the two curves before and after fusion under different strain energy densities are listed in Table 19. The predicted life of the curve after fusion is larger at a low strain energy density level. With the change in data standard deviation, the curve before and after fusion gradually deviates from high-strain energy density to low-strain energy density. The predicted life of the curve after fusion gradually exceeds that of the curve before fusion. Therefore, the standard deviation is reduced at the low strain energy density level, which is closer to the real population data. A more accurate P-W_t-N_f curve can be obtained by data fusion.

4. Conclusions

In this paper, to accurately evaluate the probabilistic characteristics of fatigue properties with small sample data under different stress ratios, a data fusion method for torsional fatigue life under different stress ratios was proposed based on the energy method. Numerical simulation was used to calculate the fatigue strain energy density. It was verified that the W_t-N_f model under different stress ratios can be represented by the same. Combined with the data fusion method, we realized data fusion for fatigue life under different stress ratios. The following conclusions were obtained:

(1) Torsional fatigue tests were conducted, and test data were obtained. Based on numerical simulation, a method for the calculation of strain energy density was established. And the strain energy densities of specimens under different working conditions were calculated to provide a database for subsequent studies.
(2) The W_t-N_f and P-W_t-N_f prediction models under two different stress ratios were fitted. By error analysis, more than 85% of the data were distributed within the ±2 scatter bands, which verified the homogeneity of the models under different stress ratios.
(3) A data fusion method for fatigue life under different stress ratios was proposed. The data fusion method can expand small sample data, significantly reducing the standard deviation of medium and low strain energy density levels. Compared with the pre-fusion data, the standard deviation of the post-fusion data was reduced by a maximum of 21.5% for the smooth specimens and 38.5% for the notched specimens. The life prediction accuracy of the fused P-W_t-N_f curves increased.

Author Contributions: Conceptualization, J.Y.; methodology, C.W.; validation, X.Z. and X.L.; formal analysis, Y.W. (Yulin Wu) and H.L.; investigation, Y.W. (Yiheng Wei); resources, J.X.; writing—original draft, C.W.; writing—review and editing, J.Y. and X.Z.; visualization, Y.W. (Yulin Wu) and X.L.; supervision, H.L. and J.X. All authors have read and agreed to the published version of the manuscript.

Funding: This research was funded by the National Science and Technology Major Project (Grant No. J2022-IV-0010-0024), and Fundamental Research Funds for the Central Universities (Grant No. 2023CDJXY-007).

Institutional Review Board Statement: Not applicable.

Informed Consent Statement: Not applicable.

Data Availability Statement: The original contributions presented in the study are included in the article, further inquiries can be directed to the corresponding author.

Conflicts of Interest: The authors declare no conflicts of interest.

Nomenclature

S	fatigue stress	n'	cyclic strain hardening index
N	fatigue life	K'	cyclic strength coefficient
P	survival rate	T	temperature
W_t	strain energy density	Kt	stress concentration factor
ΔW_e	elastic strain energy density	R	stress ratio
ΔW_p	plastic strain energy density	v	Poisson ratio
ε	strain	x_a	average fatigue life
σ	stress	x_p	probabilistic fatigue life
σ_{max}	Maximum stress	$k_{(p,\gamma,v)}$	one-side tolerance factor
$\Delta\varepsilon$	strain range	n_{ji}	fatigue life of specimen i at strain energy density level j
$\Delta\sigma$	stress range	$p\left(n_{ji}\right)$	probability of life being less than n_{ji} at strain energy density level j
ε_p	plastic strain	μ_i	Mean fatigue life at strain energy density level i
ε_e	elastic strain	σ_i	standard deviation at strain energy density level i
E	elastic modulus		

References

1. Murakami, Y.; Yokoyama, N.N.; Nagata, J. Mechanism of fatigue failure in ultralong life regime. *Fatigue Fract. Eng. Mater. Struct.* **2002**, *25*, 735–746. [CrossRef]
2. Kocańda, S. *Fatigue Failure of Metals*; Springer: Dordrecht, The Netherlands, 1978.
3. Li, W.; Yan, Q.; Xue, J. Analysis of a crankshaft fatigue failure. *Eng. Fail. Anal.* **2015**, *55*, 139–147. [CrossRef]
4. Bhaumik, S.; Sujata, M.; Venkataswamy, M. Fatigue failure of aircraft components. *Eng. Fail. Anal.* **2008**, *15*, 675–694. [CrossRef]
5. Yang, J.; Gong, Y.; Jiang, L.; Lin, W.; Liu, H. A multi-axial and high-cycle fatigue life prediction model based on critical plane criterion. *J. Mater. Res. Technol.* **2022**, *18*, 4549–4563. [CrossRef]
6. Jiang, L.; Zhang, Y.; Gong, Y.; Li, W.; Ren, S.; Liu, H. A new model characterizing the fatigue delamination growth in DCB laminates with combined effects of fiber bridging and stress ratio. *Compos. Struct.* **2021**, *268*, 113943. [CrossRef]
7. Lee, Y.L.; Barkey, M.E.; Kang, H.T. *Metal Fatigue Analysis Handbook: Practical Problem-Solving Techniques for Computer-Aided Engineering*; Elsevier: Amsterdam, The Netherlands, 2011.
8. Shao, Y.; Lu, P.; Wang, B.; Xiang, Q. Fatigue reliability assessment of small sample excavator working devices based on Bootstrap method. *Frat. Ed Integrità Strutt.* **2019**, *13*, 757–767. [CrossRef]
9. Nie, T. Application of small sample analysis in life estimation of aeroengine components. *J. Southwest Jiaotong Univ.* **2010**, *18*, 285–288.
10. Xie, L.; Liu, J.; Wu, N.; Qian, W. Probabilistic Specimen Property-Fatigue Life Mapping and P-S-N Curve Fitting. *Int. J. Reliab. Qual. Saf. Eng.* **2013**, *20*, 1350020. [CrossRef]
11. Xie, L.; Liu, J.; Wu, N.; Qian, W. Backwards statistical inference method for P–S–N curve fitting with small-sample experiment data. *Int. J. Fatigue* **2014**, *63*, 62–67. [CrossRef]
12. Chen, J.; Liu, S.; Zhang, W.; Liu, Y. Uncertainty quantification of fatigue S-N curves with sparse data using hierarchical Bayesian data augmentation. *Int. J. Fatigue* **2020**, *134*, 105511. [CrossRef]
13. Liu, X.-W.; Lu, D.-G.; Hoogenboom, P.C. Hierarchical Bayesian fatigue data analysis. *Int. J. Fatigue* **2017**, *100*, 418–428. [CrossRef]
14. Gao, J.; Yuan, Y. Small sample test approach for obtaining *P-S-N* curves based on a unified mathematical model. *Proc. Inst. Mech. Eng. Part C J. Mech. Eng. Sci.* **2020**, *234*, 4751–4760. [CrossRef]
15. Li, C.; Wu, S.; Zhang, J.; Xie, L.; Zhang, Y. Determination of the fatigue *P-S-N* curves—A critical review and improved backward statistical inference method. *Int. J. Fatigue* **2020**, *139*, 105789. [CrossRef]
16. Klemenc, J.; Fajdiga, M. Estimating S–N curves and their scatter using a differential ant-stigmergy algorithm. *Int. J. Fatigue* **2012**, *43*, 90–97. [CrossRef]
17. Shimizu, S.; Tosha, K.; Tsuchiya, K. New data analysis of probabilistic stress-life (P–S–N) curve and its application for structural materials. *Int. J. Fatigue* **2010**, *32*, 565–575. [CrossRef]
18. Bell, J. *The Experimental Foundations of Solid Mechanics*; Spinger: New York, NY, USA, 1973; pp. 5–120.
19. Inglis, N.P. Hysteresis and fatigue of Wohler rotating cantilever specimen. *Metallurgist* **1927**, *1*, 23–27.
20. Miner, M.A. Cumulative damage in fatigue. *J. Appl. Mech.* **1945**, *12*, 159–164. [CrossRef]

21. Lazzarin, P.; Zambardi, R. A finite-volume-energy based approach to predict the static and fatigue behavior of components with sharp V-shaped notches. *Int. J. Fract.* **2001**, *112*, 275–298. [CrossRef]
22. Branco, R.; Prates, P.; Costa, J.; Borrego, L.; Berto, F.; Kotousov, A.; Antunes, F. Rapid assessment of multiaxial fatigue lifetime in notched components using an averaged strain energy density approach. *Int. J. Fatigue* **2019**, *124*, 89–98. [CrossRef]
23. Braccesi, C.; Morettini, G.; Cianetti, F.; Palmieri, M. Development of a new simple energy method for life prediction in multiaxial fatigue. *Int. J. Fatigue* **2018**, *112*, 1–8. [CrossRef]
24. McCartney, L. Energy methods for fatigue damage modelling of laminates. *Compos. Sci. Technol.* **2008**, *68*, 2601–2615. [CrossRef]
25. Cao, X.; Tang, X.; Chen, L.; Wang, D.; Jiang, Y. Study on Characteristics of Failure and Energy Evolution of Different Moisture-Containing Soft Rocks under Cyclic Disturbance Loading. *Materials* **2024**, *17*, 1770. [CrossRef] [PubMed]
26. Tavernelli, J.F.; Coffi, L.F., Jr. Experimental support for generalized equation predicting low cycle fatigue. *J. Basic Eng.* **1962**, *84*, 533. [CrossRef]
27. Manson, S.S. Fatigue: A complex subject—Some simple approximations. *Exp. Mech.* **1965**, *5*, 193–226. [CrossRef]
28. Hu, Z.; Berto, F.; Hong, Y.; Susmel, L. Comparison of TCD and SED methods in fatigue lifetime assessment. *Int. J. Fatigue* **2019**, *123*, 105–134. [CrossRef]
29. Fan, Y.-N.; Shi, H.-J.; Tokuda, K. A generalized hysteresis energy method for fatigue and creep-fatigue life prediction of 316L(N). *Mater. Sci. Eng. A* **2015**, *625*, 205–212. [CrossRef]
30. Hwang, J.H.; Kim, D.W.; Lim, J.Y.; Hong, S.G. Energy-Based Unified Models for Predicting the Fatigue Life Behaviors of Austenitic Steels and Welded Joints in Ultra-Supercritical Power Plants. *Materials* **2024**, *17*, 2186. [CrossRef] [PubMed]
31. Liao, D.; Zhu, S.-P. Energy field intensity approach for notch fatigue analysis. *Int. J. Fatigue* **2019**, *127*, 190–202. [CrossRef]
32. Song, W.; Liu, X.; Zhou, G.; Wei, S.; Shi, D.; He, M.; Berto, F. Notch energy-based low and high cycle fatigue assessment of load-carrying cruciform welded joints considering the strength mismatch. *Int. J. Fatigue* **2021**, *151*, 106410. [CrossRef]
33. Berto, F. Fatigue and fracture assessment of notched components by means of the Strain Energy Density. *Eng. Fract. Mech.* **2016**, *167*, 176–187. [CrossRef]
34. El Kadi, H.; Ellyin, F. Effect of stress ratio on the fatigue of unidirectional glass fiber/epoxy composite laminae. *Composites* **1994**, *25*, 917–924. [CrossRef]
35. Kujawski, D.; Ellyin, F. A unified approach to mean stress effect on fatigue threshold conditions. *Int. J. Fatigue* **1995**, *17*, 101–106. [CrossRef]
36. Ellyin, F. *Fatigue Damage, Crack Growth and Life Prediction*; Springer: Berlin/Heidelberg, Germany, 2012.
37. Skelton, R.P. Cyclic stress-strain properties during high strain fatigue. In *High Temperature Fatigue: Properties and Prediction*; Springer: Dordrecht, The Netherlands, 1987; pp. 27–112.
38. Zhu, S.-P.; Huang, H.-Z.; He, L.-P.; Liu, Y.; Wang, Z. A generalized energy-based fatigue–creep damage parameter for life prediction of turbine disk alloys. *Eng. Fract. Mech.* **2012**, *90*, 89–100. [CrossRef]
39. Li, X.K.; Chen, S.; Zhu, S.P.; Liao, D.; Gao, J.W. Probabilistic fatigue life prediction of notched components using strain energy density approach. *Eng. Fail. Anal.* **2021**, *124*, 105375. [CrossRef]
40. Klesnil, M.; Lukác, P. *Fatigue of Metallic Materials*; Elsevier: Amsterdam, The Netherlands, 1992.
41. Branco, R.; Costa, J.D.; Berto, F.; Antunes, F.V. Fatigue life assessment of notched round bars under multiaxial loading based on the total strain energy density approach. *Theor. Appl. Fract. Mech.* **2018**, *97*, 340–348. [CrossRef]
42. Zhang, S.; Zhao, D. (Eds.) *Aerospace Materials Handbook*; CrC Press: Boca Raton, FL, USA, 2012.
43. Golos, K.; Ellyin, F. A Total Strain Energy Density Theory for Cumulative Fatigue Damage. *J. Press. Vessel. Technol.* **1988**, *110*, 36–41. [CrossRef]
44. Sun, J.; Yang, Z.; Chen, G. Research on three-parameter power function equivalent energy method for high temperature strain fatigue. In Proceedings of the 2010 the 2nd International Conference on Industrial Mechatronics and Automation, Wuhan, China, 30–31 May 2010; Volume 1, pp. 84–87.
45. Makepeace, C.E.; Ailor, W.H. Statistical Design of Experiments. In *Handbook on Corrosion Testing and Evaluation*; Jogn Wiley & Sons Inc.: Hoboken, NJ, USA, 1971.
46. Tanaka, S.; Ichikawa, M.; Akita, S. A probabilistic investigation of fatigue life and cumulative cycle ratio. *Eng. Fract. Mech.* **1984**, *20*, 501–513. [CrossRef]

materials

MDPI

Article

Study on the Constitutive Modeling of (2.5 vol%TiB + 2.5 vol%TiC)/TC4 Composites under Hot Compression Conditions

Kehao Qiang [1], Shisong Wang [2], Haowen Wang [3], Zhulin Zeng [3] and Liangzhao Qi [2,*]

[1] College of Materials and Chemistry and Chemical Engineering, Chengdu University of Technology, Chengdu 610059, China; q1430347908@outlook.com
[2] College of Nuclear Technology and Automation Engineering, Chengdu University of Technology, Chengdu 610059, China; wsscdp@outlook.com
[3] School of Mechanical and Electrical Engineering, Chengdu University of Technology, Chengdu 610059, China; wdyx_teq@outlook.com (H.W.); zengzhulin1212@outlook.com (Z.Z.)
* Correspondence: qiliangzhao18@cdut.edu.cn

Abstract: The hot deformation behavior of titanium matrix composites plays a crucial role in determining the performance of the formed components. Therefore, it is significant to establish an accurate constitutive relationship between material deformation parameters and flow stress. In this study, hot compression experiments were conducted on a (2.5 vol%TiB + 2.5 vol%TiC)/TC4. The experiments were performed under temperatures ranging from 1013.15 to 1133.15 K and strain rates ranging from 0.001 to 0.1 s^{-1}. Based on the stress–strain data obtained from the experiment, the constitutive models were established by using the Arrhenius model and the BP neural network algorithm, respectively. Considering the relationship between strain rate, hot working temperature, and flow stress, a comparative analysis was conducted to evaluate the prediction accuracy of two different constitutive models. The research results indicate that the flow stress of (2.5 vol%TiB + 2.5 vol%TiC)/TC4 increases with decreasing temperature and increasing strain rate, and the stress–strain curve shows obvious work hardening and softening behaviors. Both the Arrhenius model and the BP neural network algorithm are effective in predicting the hot compression flow stress of (2.5 vol%TiB + 2.5 vol%TiC)/TC4, but the average relative error and root mean square error of the BP neural network algorithm are smaller and the correlation coefficient is higher, thus possessing higher accuracy and reliability.

Keywords: titanium matrix composites; hot working; flow stress; constitutive model

1. Introduction

Titanium matrix composites (TMCs) are extensively utilized in the aerospace and automotive industries due to their exceptional corrosion resistance, high temperature resistance, and mechanical properties [1–3]. (TiB + TiC)/TC4 is a typical TMC, and relevant studies [4] have indicated that the composite has the best combination of strength, modulus, and plasticity by introducing TiB and TiC simultaneously. Components made from (TiB + TiC)/TC4 typically undergo complicated casting and forging processes where the hot deformation process impacts the microstructural evolution and flow stress behavior of the material, consequently affecting the mechanical properties of the final formed component [5–8]. Therefore, it is essential to comprehend the influence of hot deformation parameters on microstructure evolution and flow stress for optimizing the hot working technology.

In the hot deformation process of metallic materials, the strain rate and hot working temperature are important factors that affect the material deformation behavior. Li et al. [9] investigated the deformation behavior of the TC18 alloy under various strain rates and hot working temperatures, analyzing the deformation behavior of the α and β phases. Cui et al. [10] conducted hot compression experiments to analyze the impact of different hot working conditions on flow stress of the TB6 alloy. Dan et al. [11] investigated the hot deformation behavior of the TC11 alloy and pointed out that the microstructure evolution under low strain rates was dominated by lamellar tissue spheroidization. Furthermore, several studies [12–14] have indicated the substantial influence of strain rate and hot working temperature on microstructure evolution and flow stress. Consequently, it is imperative to predict the hot deformation behavior of TMCs accurately under different strain rates and hot working temperatures.

Constitutive modeling is an effective mathematical modeling method for reflecting material flow stress behavior that has been widely applied in metallic materials [15–17]. According to the intrinsic mechanism of the model, constitutive models can be classified into phenomenological models, empirical models, and deep learning algorithm models. While phenomenological models have inherent physical meanings, empirical models and deep learning algorithm models are simpler and more suitable for intuitively analyzing the influence of strain rate and hot working temperature on the flow stress of metallic materials [18–20]. For example, Deb et al. [21] utilized a modified J-C model and deep learning algorithm to develop a constitutive model for TC4 alloy, which effectively predicted the flow stress behavior under various hot working conditions. Yang et al. [22] used the Arrhenius model and the BP neural network algorithm to accurately describe the constitutive relationship between strain rate, hot working temperature, and flow stress of the 7075 alloy, and they analyzed the prediction accuracy of these two models. Yu et al. [23] utilized empirical models and deep learning algorithm models to develop multiple sets of constitutive models for the TG6 alloy, and they compared the applicability of different models. The above studies indicate that both empirical models and deep learning algorithm models are effective in predicting the flow stress of metallic materials under different hot working conditions.

In this study, constitutive models for (2.5 vol%TiB + 2.5 vol%TiC)/TC4 were established based on the Arrhenius model and BP neural network algorithm, which aim to accurately predict the relationship between strain rate, hot working temperature, and flow stress. The subsequent research contents are as follows: Section 2 plots the stress–strain curves of (2.5 vol%TiB + 2.5 vol%TiC)/TC4 through hot compression experiments and analyzes the deformation behavior. Section 3 presents the constitutive models considering strain rate, hot deformation temperature, and flow stress of the TMCs by using the Arrhenius model and BP neural network algorithm, respectively. Section 4 compares the prediction accuracy of two models. Section 5 provides a summary and conclusions.

2. Experimental Methods and Results

The TMC used in this study is (2.5 vol%TiB + 2.5 vol%TiC)/TC4; both TiB and TiC particles have a volume fraction of 2.5%. The initial microstructure of the sample is shown in Figure 1a. The matrix material used in this study is TC4 titanium alloy, which consists of an α phase and an intergranular β phase. The second-phase particles are TiB and TiC, in which TiB has a lamellar shape and TiC has an equiaxial shape. Figure 1b illustrates the specific hot compression experiment, which is divided into four stages according to the relevant references [5–7]. Firstly, the specimen is heated from the normal temperature to the hot compression temperature. Subsequently, keep the specimen warm for 5 min to avoid temperature deviations. In the third stage, the specimen is compressed from the original height (15 mm) to 50% (7.5 mm) with a specified strain rate, corresponding to a true strain of 0.7. Finally, the specimen is cooled to a normal temperature in the air for observation. The hot working temperature and strain rate conditions for the experiment are listed in Table 1. The stress–strain behavior of the TMCs in the two-phase region is studied in this paper. Therefore, the hot working temperature should be less than the temperature of the phase transition point [24].

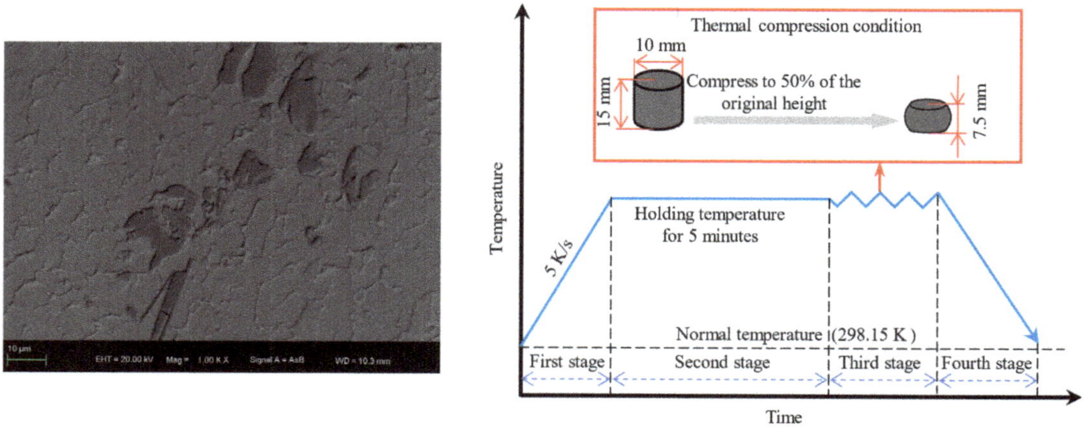

(**a**) (**b**)

Figure 1. Experiment methods: (**a**) initial microstructure; (**b**) experimental process.

Table 1. Experiment conditions.

Serial Number	Strain Rate (s^{-1})	Temperature (K)
1	0.001	
2	0.01	1013.15
3	0.1	
4	0.001	
5	0.01	1053.15
6	0.1	
7	0.001	
8	0.01	1093.15
9	0.1	
10	0.001	
11	0.01	1133.15
12	0.1	

The Gleeble 3500 was used for the hot compression experiment. To ensure the validity of the experiment results, two experiments were conducted under each experiment condition, and the stress–strain curve error of the two experiments was ensured to be less than 5%. Meanwhile, according to [25,26], the influence of temperature and friction coefficient on the stress–strain curve was modified. The true stress–strain curves of the specimen under different strain rates and temperatures are shown in Figure 2. It shows that with the temperature increasing and the strain rate decreasing, the flow stress decreases. Additionally, the flow stress curves under all conditions have work hardening, dynamic recovery (DRV), and dynamic recrystallization (DRX) behaviors. In the initial deformation stage, a significant amount of dislocation proliferation and accumulation occurs, resulting in a rapid increase and peak in flow stress. This behavior is known as work hardening. As strain increases, the softening effect caused by the DRV process strengthens gradually. Simultaneously, internal energy accumulates in the composites, leading to the nucleation and growth of recrystallized grains, which triggers the DRX process and further enhances the softening effect. As the DRV and DRX processes progress, the work hardening effect gradually weakens, resulting in a gradual decrease in flow stress and a tendency towards stabilization (Note: some data in Figure 2 have been used in our previous study [27]).

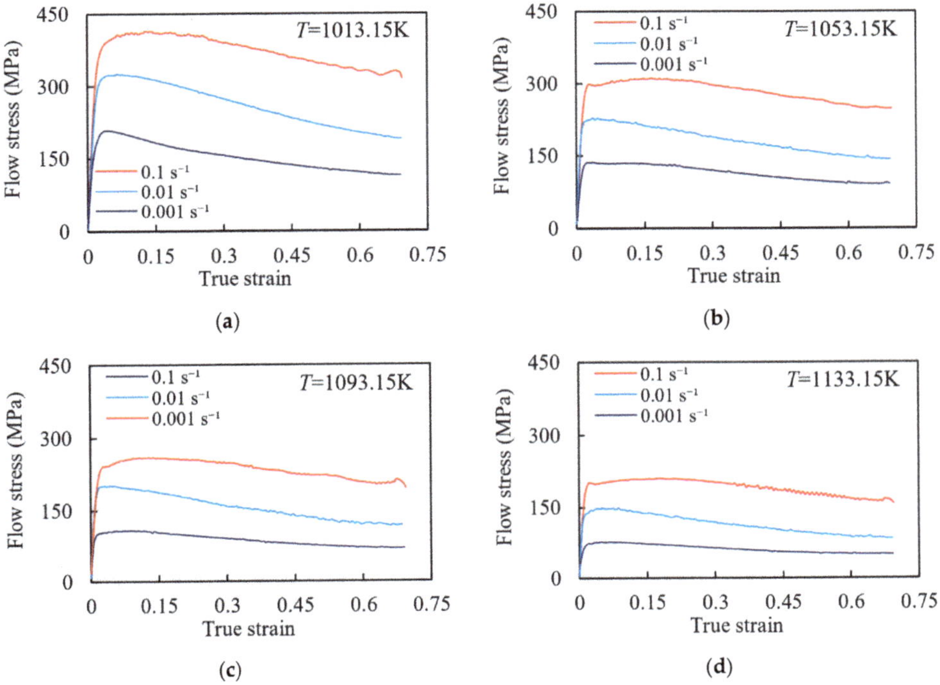

Figure 2. Stress–strain curves under different temperatures: (**a**) 1013.15 K; (**b**) 1053.15 K; (**c**) 1093.15 K; (**d**) 1133.15 K.

The results of electron back scatter diffraction (EBSD) under different strain rates are shown in Figure 3. It can be seen that as the strain rate decreases, the fraction and grain sizes of DRX increase significantly. This is because the higher strain rate provides a greater driving force for the DRX process, resulting in a faster DRX nucleation rate. Simultaneously, the high strain rate hinders the growth of DRX grains, forming refined equiaxed DRX grains. Therefore, a decrease in strain rate leads to an increase in both the fraction and grain size of DRX.

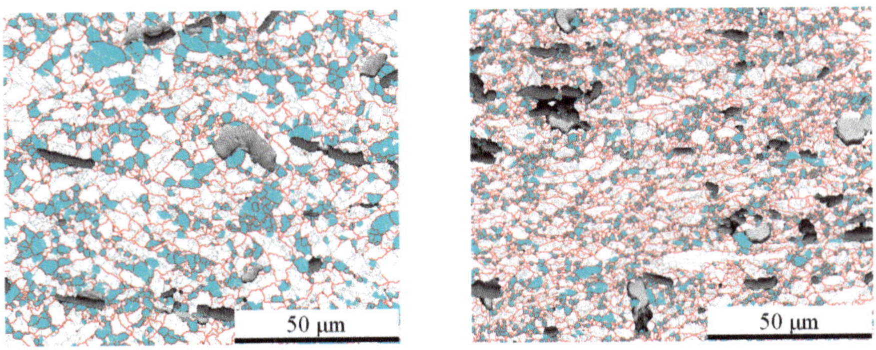

(**a**) (**b**)

Figure 3. *Cont.*

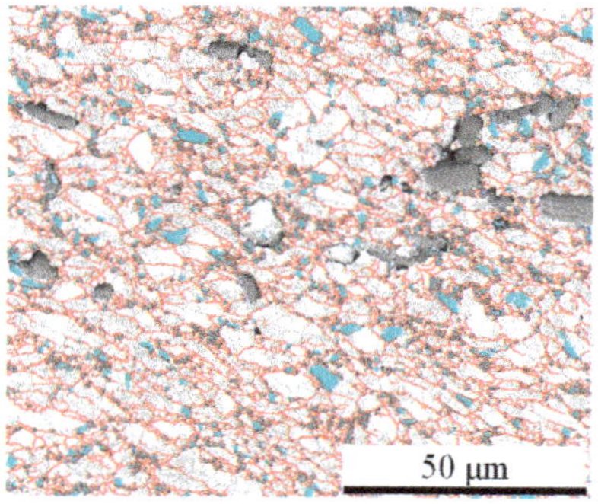

(**c**)

Figure 3. EBSD analysis results under the temperature of 1053.15 K and different strain rates: (**a**) 0.001 s^{-1}, (**b**) 0.01 s^{-1}, and (**c**) 0.1 s^{-1}.

Figure 4 shows the EBSD results under different temperatures. Contrary to the effect of strain rate, an increase in temperature leads to an increase in the DRX fraction. This is due to the nucleation and growth processes of DRX grains being influenced by thermodynamics and higher temperatures promoting dislocation evolution and grain boundary migration. Consequently, this leads to a significant increase in the nucleation rate and size of DRX grains.

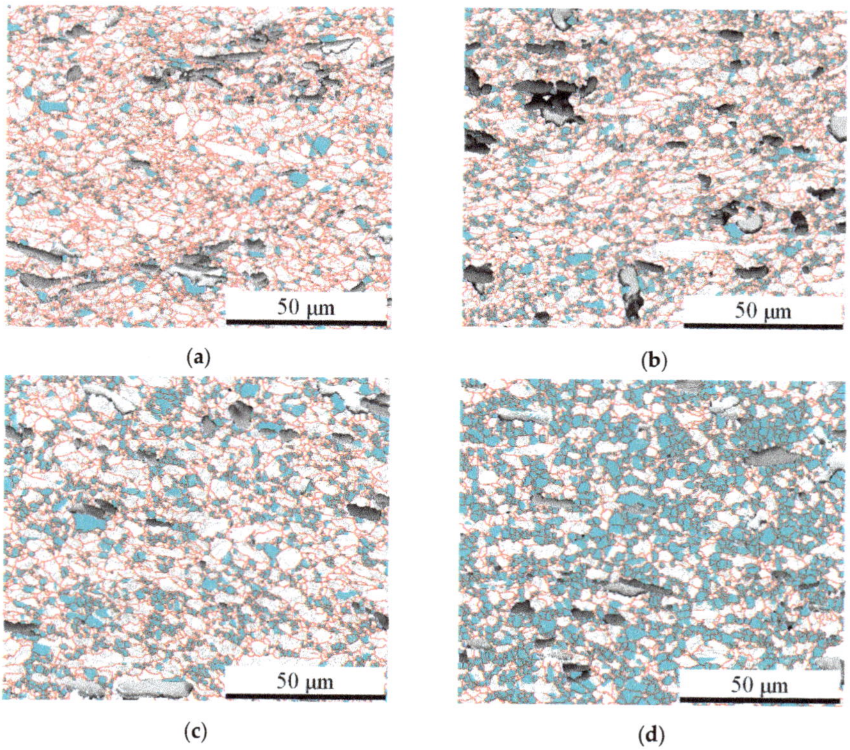

Figure 4. EBSD analysis results under the strain rate of 0.01 s^{-1} and different temperatures: (**a**) 1013.15 K, (**b**) 1053.15 K, (**c**) 1093.15 K, and (**d**) 1133.15 K.

3. Description of Constitutive Model

In this section, the Arrhenius model and the BP neural network algorithm were used to establish constitutive models for (2.5 vol%TiB + 2.5 vol%TiC)/TC4. The two constitutive models were compiled in MATLAB R2022b software to predict the flow stress under different temperatures, strain rates, and strains.

3.1. Arrhenius Model

The Arrhenius model, proposed by Sellars and Tegart [28], is an empirical constitutive model commonly used to describe the relationship between flow stress, strain rate, and deformation temperature during the hot deformation process of metal materials. It is widely utilized due to the minimal number of parameters and high accuracy. The mathematical expression for the Arrhenius model is as follows:

$$\dot{\varepsilon} = AF(\sigma)\exp\left(-\frac{Q}{RT}\right) \tag{1}$$

At different stress levels, $F(\sigma)$ can be expressed by Equation (2):

$$F(\sigma) = \begin{cases} \sigma^n & \alpha\sigma < 0.8 \\ \exp(\beta\sigma) & \alpha\sigma > 1.2 \\ [\sinh(\alpha\sigma)]^m & \text{all } \sigma \end{cases} \tag{2}$$

where: σ, ε, and T represent the flow stress (MPa), strain rate (s^{-1}), and hot working temperature (K), respectively; R and Q are the gas constant (8.314 J/(mol·K)) and deformation activation energy (kJ/mol), respectively; A, α, and β are material constants; m and n are the work hardening exponent; and $\alpha = \beta/n$.

The following can be derived from Equations (1) and (2) by applying logarithms to both sides:

$$\ln\dot{\varepsilon} = \ln A_1 + n\ln\sigma - \frac{Q}{RT} \tag{3}$$

$$\ln\dot{\varepsilon} = \ln A_2 + \beta\sigma - \frac{Q}{RT} \tag{4}$$

$$\ln\dot{\varepsilon} = \ln A_3 - \frac{Q}{RT} + m\ln[\sinh(\alpha\sigma)] \tag{5}$$

In Equations (3) and (4), under the condition of constant temperature, a linear relationship exists between $\ln\dot{\varepsilon}$ and $\ln\sigma$ and between $\ln\dot{\varepsilon}$ and σ, whose slopes are denoted as n and β, respectively.

Taking ε = 0.1 as an example, Figure 5 shows the plots of $\ln\dot{\varepsilon}$-$\ln\sigma$ and $\ln\dot{\varepsilon}$-σ for (2.5 vol%TiB + 2.5 vol%TiC)/TC4 within the range of ε = 0.1 to 0.001 s^{-1}. By calculating the average slope, the results are as follows: n = 5.2816, β = 0.0285; further, $\alpha = \beta/n$ = 0.005396.

Under the condition that $1/T$ is constant, both sides of Equation (5) take the partial derivative of $\dot{\varepsilon}$ and Equation (6) is obtained:

$$m = \frac{\partial\ln[\sinh(\alpha\sigma)]}{\partial\ln\dot{\varepsilon}} \tag{6}$$

Under the condition that $\dot{\varepsilon}$ is constant, both sides of Equation (5) take the partial derivative of $1/T$ and Equation (7) is obtained:

$$Q = Rm\frac{\partial\ln[\sinh(\alpha\sigma)]}{\partial(1/T)} \tag{7}$$

Assuming $b = \dfrac{\partial \ln[\sinh(\alpha\sigma)]}{\partial(1/T)}$, Equation (7) can be simplified as:

$$Q = Rmb \tag{8}$$

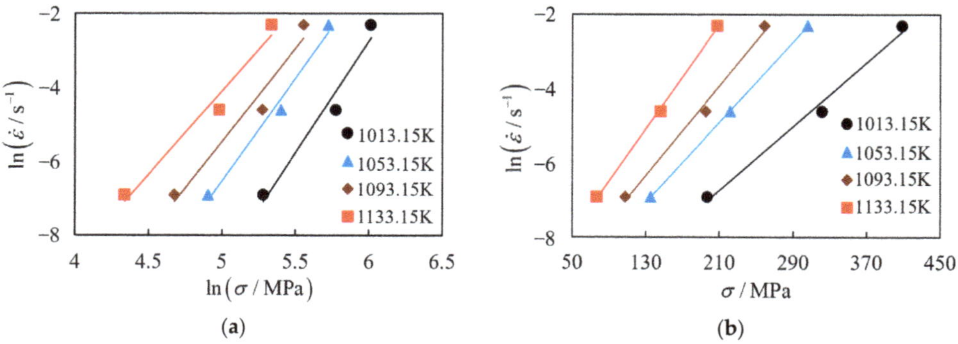

Figure 5. Linear relationship between stress and strain: (**a**) $\ln\dot{\varepsilon}$-$\ln\sigma$; (**b**) $\ln\dot{\varepsilon}$-σ.

By substituting α = 0.005396 into Equation (5), the linear relationship between $\ln\dot{\varepsilon}$-$\ln[\sinh(\alpha\sigma)]$ and $\ln[\sinh(\alpha\sigma)]$-$1000/T$ under constant temperature and constant stress can be obtained, whose slopes are expressed as m and b, respectively. Taking ε = 0.1 as an example, Figure 6 shows the plots of $\ln\dot{\varepsilon}$-$\ln[\sinh(\alpha\sigma)]$ and $\ln[\sinh(\alpha\sigma)]$-$1000/T$ for (2.5 vol%TiB + 2.5 vol%TiC)/TC4 composites within the range of ε = 0.1–0.001 s^{-1}. By calculating the average slope, the results are as follows: m = 3.8622, b = 10.5647, and Q = 339.2360 kJ/mol.

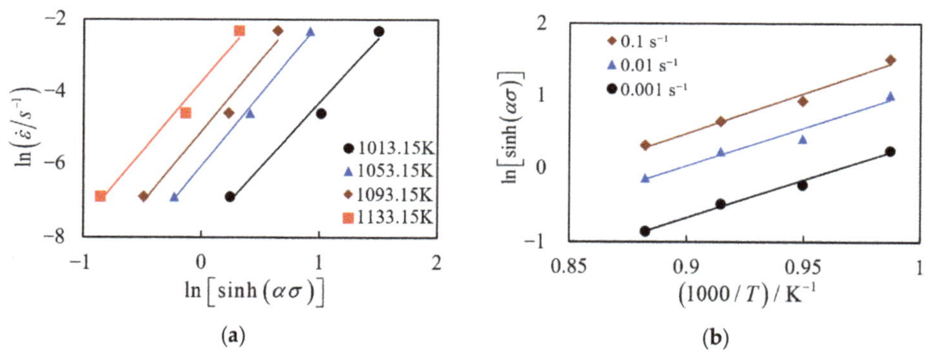

Figure 6. Linear relationship of stress–strain rate and stress–temperature: (**a**) $\ln\dot{\varepsilon}$-$\ln[\sinh(\alpha\sigma)]$; (**b**) $\ln[\sinh(\alpha\sigma)]$-$1000/T$.

According to relevant research, the influence of hot deformation temperature and strain rate on the hot deformation can be described by the temperature compensation deformation rate factor Z [29], which is expressed as:

$$Z = \dot{\varepsilon}\exp\left(-\frac{Q}{RT}\right) = A[\sinh(\alpha\sigma)]^n \tag{9}$$

Both sides of Equation (9) take the logarithms to obtain Equation (10):

$$\ln Z = \ln A + n\ln[\sinh(\alpha\sigma)] \tag{10}$$

As shown in Figure 7, a linear relationship between $\ln Z$ and $\ln[\sinh(\alpha\sigma)]$ can be obtained by combining the Q value, Equations (9) and (10). Obviously, the intercept in Figure 7 is $\ln A$ = 32.356.

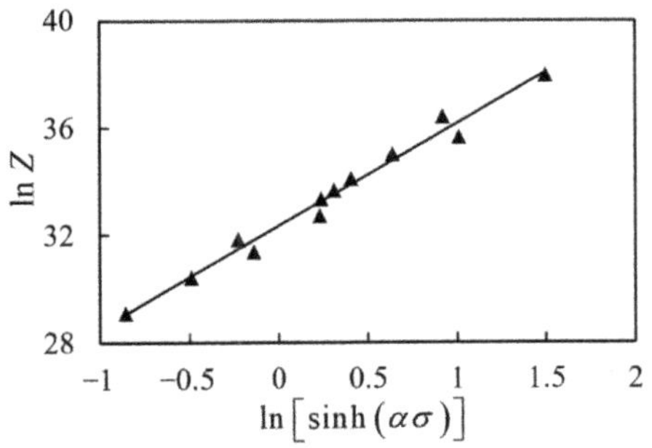

Figure 7. The linear relationship of $\ln Z$-$\ln[\sinh(\alpha\sigma)]$.

Repeating the above solving process, with ε = 0.1 as the gradient, values of α, m, Q, and $\ln A$ for (2.5 vol%TiB + 2.5 vol%TiC)/TC4 within the strain range of 0.1 to 0.7 are calculated, respectively. The results are listed in Table 2.

Table 2. Material constants of Arrhenius model.

ε	α	m	Q (kJ/mol)	ln$A$
0.1	0.005396	3.8622	339.236	32.356
0.2	0.005514	3.5735	310.272	29.334
0.3	0.005773	3.4156	294.554	27.724
0.4	0.006166	3.2785	282.275	26.409
0.5	0.006542	3.1902	270.509	25.185
0.6	0.006927	3.1987	273.668	25.604
0.7	0.007235	3.2396	272.444	25.448

Figure 8 shows the data for α, m, Q, and $\ln A$ under different strains. The change pattern between these parameters and strain is represented by a curve obtained through fifth-order polynomial fitting [30], and the specific expression is given by Equations (11)–(14). Figure 8 clearly illustrates that the fitted curve accurately reflects the variation law of each parameter and strain.

$$\alpha = 0.013\varepsilon^5 - 0.022\varepsilon^4 + 0.003\varepsilon^3 + 0.01\varepsilon^2 - 0.002\varepsilon + 0.006 \quad (11)$$

$$m = -104.333\varepsilon^5 + 215.364\varepsilon^4 - 166.099\varepsilon^3 + 61.77\varepsilon^2 - 12.699\varepsilon + 4.66 \quad (12)$$

$$Q = -16917.083\varepsilon^5 + 34067.992\varepsilon^4 - 25638.434\varepsilon^3 + 9144.104\varepsilon^2 - 1701.418\varepsilon + 440.404 \quad (13)$$

$$\ln A = -1951.25\varepsilon^5 + 3910.417\varepsilon^4 - 2924.771\varepsilon^3 + 1033.121\varepsilon^2 - 188.512\varepsilon + 43.436 \quad (14)$$

Furthermore, by rearranging Equations (1) and (2), the constitutive relationship for (2.5 vol%TiB + 2.5 vol%TiC)/TC4 within the range of hot working temperature from 1013.15 to 1133.15 K and strain rate from 0.001 to 0.1 s^{-1} can be expressed as:

$$\sigma = \frac{1}{\alpha}\ln\left\{\left(\frac{\dot{\varepsilon}}{A}\exp\left(\frac{Q}{RT}\right)\right)^{1/m} + \left[\left(\frac{\dot{\varepsilon}}{A}\exp\left(\frac{Q}{RT}\right)\right)^{2/m} + 1\right]^{1/2}\right\} \quad (15)$$

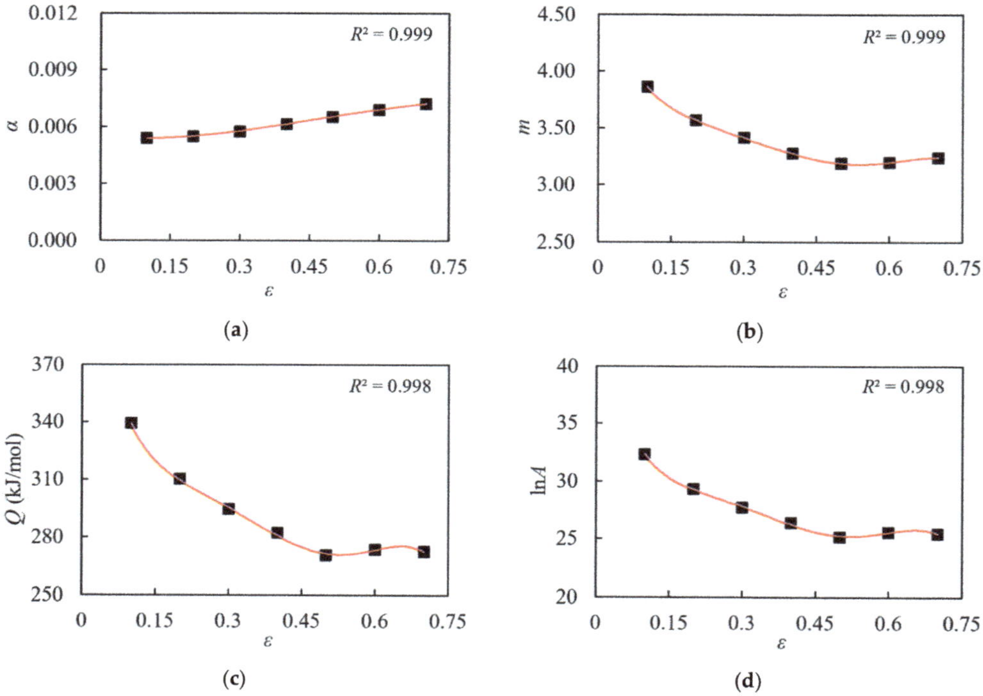

Figure 8. Fitting curves of different parameters with strain: (**a**) α-ε; (**b**) m-ε; (**c**) Q-ε; (**d**) ln A-ε.

The α, m, Q, and A values under various strain levels are calculated using Equations (11)–(14) and then incorporated into Equation (15), from which the constitutive relationship between flow stress and strain in the (2.5 vol%TiB + 2.5 vol%TiC)/TC4 under different strain rates and hot working temperatures is obtained. As shown in Figure 9, the results obtained from the Arrhenius model are in good agreement with the experimental results under various temperatures and strain rates. It is suggested that the Arrhenius model is capable of accurately predicting the flow stress behavior of the (2.5 vol%TiB + 2.5 vol%TiC)/TC4 during the hot working process.

3.2. BP Neural Network Algorithm

In the process of establishing empirical or semi-empirical constitutive models for materials using the Arrhenius equation, numerical errors are introduced by numerical methods such as approximate solutions and linear regression fitting [31]. Therefore, deep learning algorithms have been increasingly utilized in constitutive modeling due to their non-linear modeling capabilities, adaptability, and generalization abilities [32,33]. The BP neural network algorithm, proposed by Rumelhart and McClelland [34], is widely recognized and extensively used. The BP neural network algorithm model comprises two processes: signal forward propagation and error back propagation, in which the weights and thresholds are adjusted iteratively. The training process concludes when the pre-set model training times are reached or the error is reduced to an acceptable level [35].

As shown in Figure 10, the structure of the BP neural network algorithm in this study includes an input layer, hidden layers, and an output layer. The input layer consists of three neurons that receive external data corresponding to deformation temperature, strain rate, and strain. The hidden layer is used for nonlinear transformation and feature extraction. In order to prevent overfitting, the hidden layer is designed as a single layer with 16 neurons, which is determined based on references [36]. The output layer, responsible for the final data output, consists of one neuron representing the flow stress.

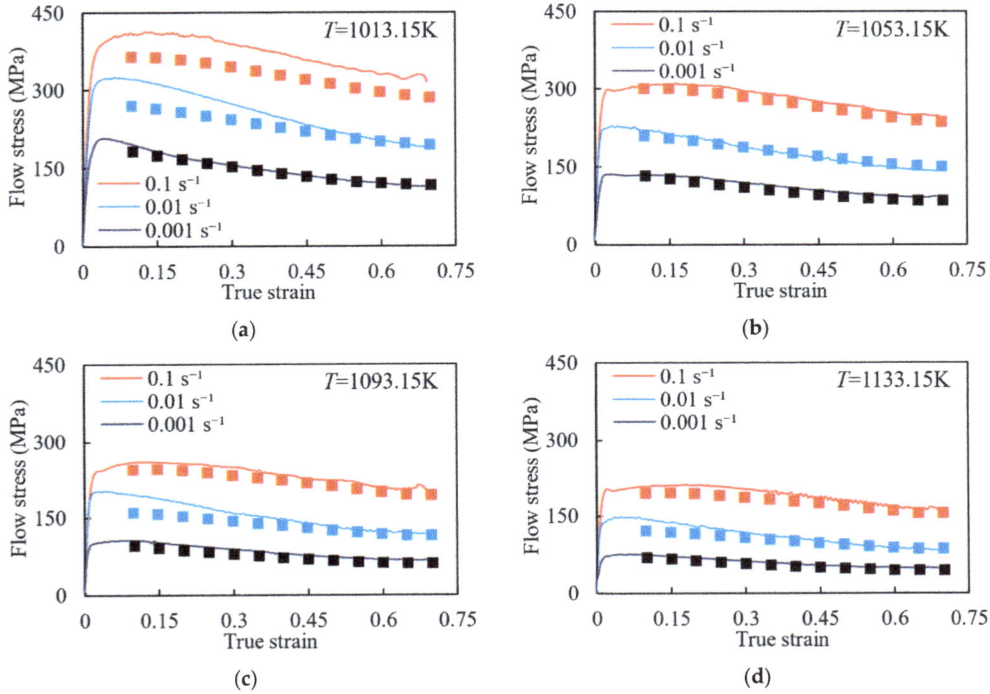

Figure 9. Comparison of stress–strain curves between experimental results (curves) and Arrhenius model prediction results (symbols): (**a**) 1013.15 K; (**b**) 1053.15 K; (**c**) 1093.15 K; and (**d**) 1133.15 K.

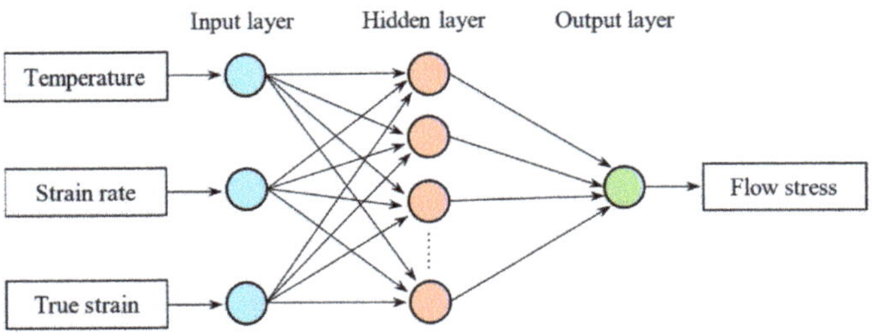

Figure 10. The BP neural network structure established in this study.

In addition, due to the significant difference in data magnitude and range between temperature, strain rate, strain, and flow stress, it is necessary to use the linear normalization method. As indicated in Equations (16) and (17), normalization and denormalization of the input and output data are necessary in order to improve the computational speed, accuracy, and robustness of the BP neural network model:

$$x' = (x - x_{\min})/(x_{\max} - x_{\min}) \tag{16}$$

$$x = x'(x_{\max} - x_{\min}) + x_{\min} \tag{17}$$

where x and x' are the data before and after normalization, respectively, and $x_{\max}$ and $x_{\min}$ are the maximum and minimum value, respectively, of the data before normalization.

The experimental data within the strain range of 0.1 to 0.7, with a gradient of 0.005, were used as data samples for a total of 1428 sets of data. Among these, data under deformation temperatures of 1013.15 K, 1053.15 K, and 1133.15 K were selected as the training set for the model, totaling 1071 sets of data; data under a temperature of 1093.15 K were used as the validation set, consisting of 357 sets of data. During the model training, the maximum number of iterations was set to 15,000, the learning rate to 0.1, and the convergence error to 0.0001. The transfer function for the hidden layer was set as the hyperbolic tangent function, while the transfer function for the output layer was set as the linear transfer function. Figure 11 shows the prediction results of the BP neural network algorithm. It is noteworthy that the stress–strain data of 1013.15 K, 1053.15 K, and 1133.15 K were used to model training, meaning the calculated results are highly consistent with the experimental results. However, the stress–strain data of 1093.15 K were only used to verify the validity of the BP neural network model instead of model training. Therefore, when analyzing the validity of this model, it is suggested to focus on Figure 11c. The BP neural network algorithm has a strong agreement with the experimental results under different hot working conditions, thereby validating the accuracy of the model established by the BP neural network algorithm.

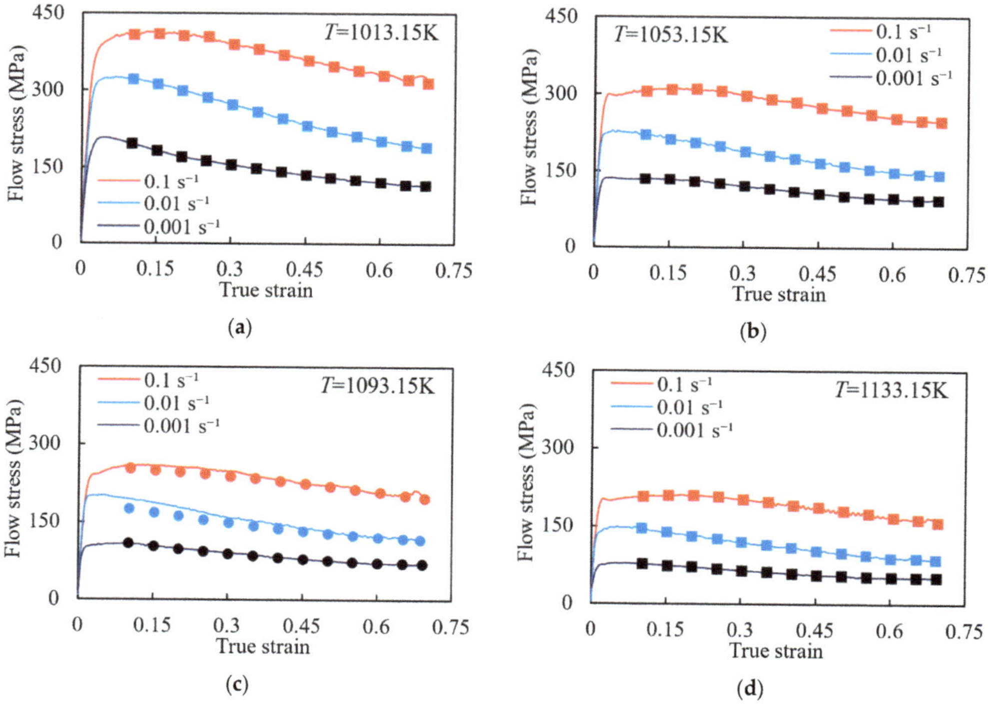

Figure 11. Comparison of stress–strain curves between experimental results (curves) and BP neural network prediction results (symbols), where 1093.15 K is the prediction result: (**a**) 1013.15 K; (**b**) 1053.15 K; (**c**) 1093.15 K; and (**d**) 1133.15 K.

4. Model Evaluation

To assess the accuracy of the above two models in predicting the hot deformation behaviors of (2.5 vol%TiB + 2.5 vol%TiC)/TC4, Equations (18)–(21) were used to calculate the relative error (*RE*), average absolute relative error (*AARE*), root mean square error

(*RMSE*), and correlation coefficient (*R*), which provide a measure of the disparity between the predicted and experimental results:

$$RE = \left| \frac{\sigma_E^i - \sigma_T^i}{\sigma_T^i} \right| \times 100\% \tag{18}$$

$$AARE = \frac{1}{N} \sum_{i=1}^{N} \left| \frac{\sigma_E^i - \sigma_T^i}{\sigma_T^i} \right| \times 100\% \tag{19}$$

$$RMSE = \sqrt{\frac{1}{N} \sum_{i=1}^{N} \left(\sigma_E^i - \sigma_T^i \right)^2} \tag{20}$$

$$R = \frac{\sum_{i=1}^{N} \left(\sigma_T^i - \overline{\sigma}_T \right) \left(\sigma_E^i - \overline{\sigma}_E \right)}{\sqrt{\sum_{i=1}^{N} \left(\sigma_T^i - \overline{\sigma}_T \right)^2} \sqrt{\sum_{i=1}^{N} \left(\sigma_E^i - \overline{\sigma}_E \right)^2}} \tag{21}$$

where σ and $\overline{\sigma}$ represent true stress and mean value of true stress, respectively, and subscripts E and T represent the predicted values and experimental values, respectively; N represents the total number of samples, and i represents the i-th item in the samples (in each working condition, 119 groups of data in the strain range of 0.1 to 0.7 were selected for calculation with a gradient of 0.05).

Figure 12 displays the relative error distributions of the Arrhenius model-predicted values under four different hot working temperatures. The *AARE* value was calculated at 6.54%; it is noteworthy that 48.32% of the *RE* values are greater than 6.54% while 51.68% are less than 6.54%. It shows that the predicted values are in good agreement with the experimental values.

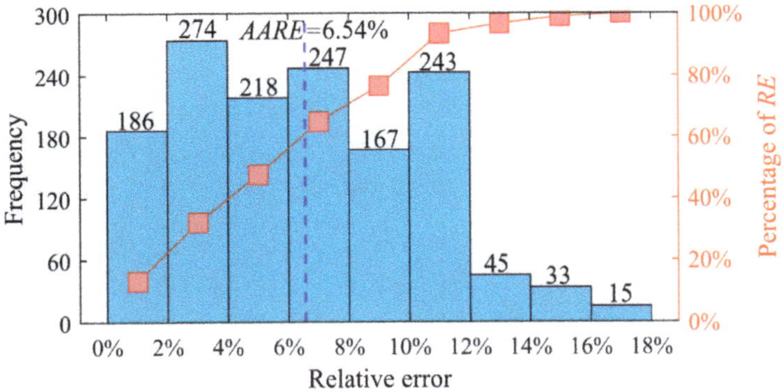

Figure 12. The relative error distribution of the Arrhenius model-predicted values.

Figure 13 shows the correlation between the predicted values and the experimental values. Most of the data fall within the mean relative error range of ±6.54%. The *RMSE* values under the four hot working temperatures are as follows: 28.225, 8.112, 12.989, and 9.678; the corresponding *R* values are 0.997, 0.996, 0.996, and 0.997, respectively. The overall *RMSE* and *R* values for the data are 16.769 and 0.994, respectively. These results indicate that the Arrhenius model has low prediction errors and significant correlations between the predicted values and experimental values. Therefore, the Arrhenius model effectively predicts the flow stress behavior of (2.5 vol%TiB + 2.5 vol%TiC)/TC4.

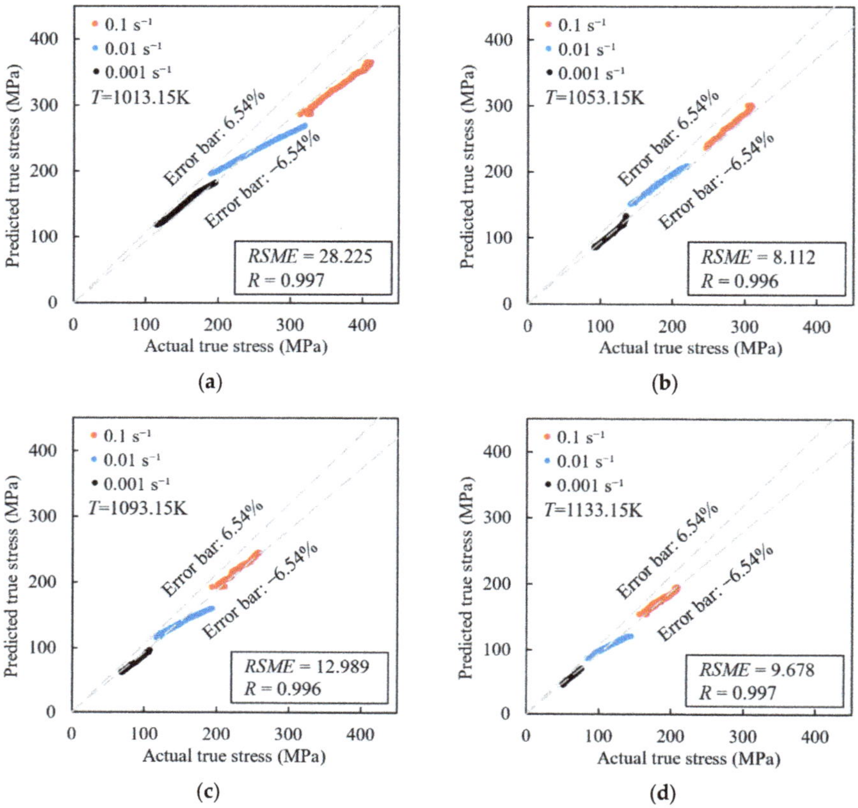

Figure 13. The correlation between Arrhenius model-predicted values and experimental values: (**a**) 1013.15 K; (**b**) 1053.15 K; (**c**) 1093.15 K; and (**d**) 1133.15 K.

Figure 14 shows the relative error distribution of the predicted values for the validation set (at 1093.15 K) of the BP neural network algorithm. The *AARE* value is 3.00%, with 36.13% of the *RE* values being greater than 3.00% and 63.87% being less than 3.00%. These findings suggest a strong correlation between the predicted values and the experimental values.

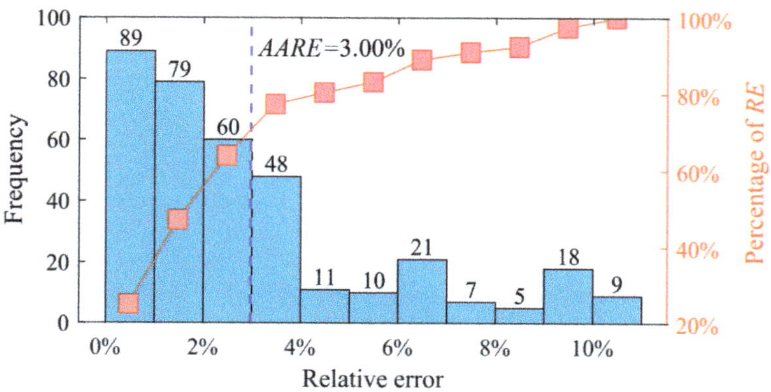

Figure 14. The relative error distribution of BP neural network algorithm prediction values.

Figure 15 shows the correlation between the predicted values and the experimental values in the validation set. It can be seen that most of the data fall within the mean relative error range of ±3.00%, and the *RMSE* and *R* values were reported as 7.311 and 0.997, respectively. These findings indicate that the BP neural network algorithm has greater accuracy in predicting the flow stress of the (2.5 vol%TiB + 2.5 vol%TiC)/TC4 in comparison to the Arrhenius model.

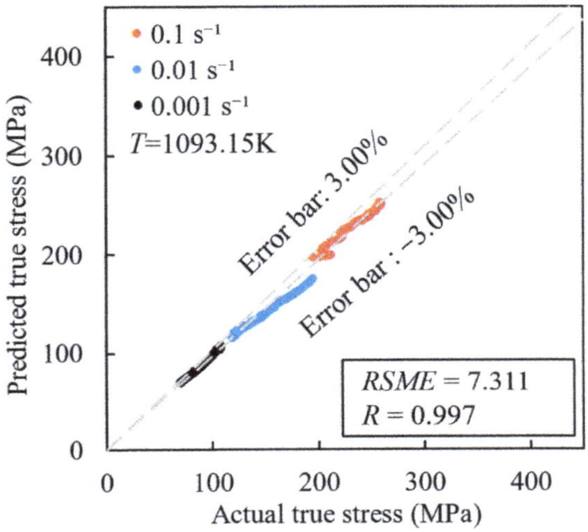

Figure 15. Correlation between BP neural network algorithm prediction values and experimental values.

5. Conclusions

In this study, the stress–strain data of (2.5 vol%TiB + 2.5 vol%TiC)/TC4 under different temperatures and strain rates were obtained by hot compression experiments. Based on the experimental results, the constitutive models of the TMCs were established by the Arrhenius model and the BP neural network algorithm, respectively. The main conclusions are the following:

(1) The flow stress of the (2.5 vol%TiB + 2.5 vol%TiC)/TC4 increases with the strain rate increasing and the hot working temperature decreasing. The hot deformation process is influenced by microstructure evolution behaviors such as dynamic recovery, dynamic globularization, and dynamic recrystallization. Additionally, the stress–strain curve has obvious work hardening and softening behaviors.

(2) Based on the hot compression experimental results, constitutive models of (2.5 vol%TiB + 2.5 vol%TiC)/TC4 were established using the Arrhenius model and the BP neural network algorithm. The average absolute relative error, root mean square error, and correlation coefficient of the Arrhenius model are 6.54%, 16.769, and 0.994, respectively. The average absolute relative error, root mean square error, and correlation coefficient of the BP neural network algorithm are 3.00%, 7.311, and 0.997, respectively.

(3) Both the Arrhenius model and the BP neural network algorithm can be used to predict the flow stress of (2.5 vol%TiB + 2.5 vol%TiC)/TC4 under different hot working conditions. However, the BP neural network algorithm offers a simpler modeling process and higher prediction accuracy, making it more suitable for accurately predicting the hot compression flow stress of TMCs.

Author Contributions: Conceptualization, K.Q. and L.Q.; methodology, K.Q. and S.W.; validation, S.W.; investigation, H.W. and Z.Z.; resources, S.W. and L.Q.; writing—original draft preparation, K.Q. and S.W.; writing—review and editing, H.W. and Z.Z.; visualization, K.Q. and H.W.; supervision, L.Q. All authors have read and agreed to the published version of the manuscript.

Funding: This research received no external funding.

Institutional Review Board Statement: Not applicable.

Informed Consent Statement: Not applicable.

Data Availability Statement: The data presented in this study are available on request from the corresponding author.

Conflicts of Interest: The authors declare no conflict of interest.

References

1. Caha, I.; Alves, A.C.; Kuroda, P.A.B.; Pinto, A.M.P.; Rocha, L.A.; Toptan, F. Degradation behavior of Ti-Nb alloys: Corrosion behavior through 21 days of immersion and tribocorrosion behavior against alumina. *Corros. Sci.* **2022**, *167*, 108488. [CrossRef]
2. Banerjee, B.; Williams, J.C. Perspectives on titanium science and technology. *Acta Mater.* **2013**, *61*, 844–879. [CrossRef]
3. Jahani, A.; Aval, H.J.; Rajabi, M.; Jamaati, R. Effects of Ti_2SnC max phase reinforcement content on the properties of copper matrix composite produced by friction stir back extrusion process. *Mater. Chem. Phys.* **2023**, *299*, 127497. [CrossRef]
4. Liu, W.-J.; Guo, X.-L.; Meng, J.-L.; Wang, F.-Q.; Wang, L.-Q.; Zhang, D. Analysis of the coupling effects of tib whiskers and tic particles on the fracture toughness of (TiB + TiC)/TC4 composites: Experiment and modeling. *Met. Mater. Trans. A* **2015**, *46*, 3490–3501. [CrossRef]
5. Sajadifar, S.V.; Maier, H.J.; Niendorf, T.; Yapici, G.G. Elevated temperature mechanical characteristics and fracture behavior of a novel beta titanium alloy. *Crystals* **2023**, *13*, 269. [CrossRef]
6. Nochovnaya, N.A.; Shiryaev, A.A.; Davydova, E.A. Influence of manufacturing parameters on pseudo-β-titanium alloy VT47 sheet structural state and mechanical property anisotropy. *Metallurgist* **2023**, *66*, 1216–1224. [CrossRef]
7. Illarionov, A.G.; Vodolazskii, F.V.; Kosmatskii, Y.I.; Gornostaeva, E.A. Effect of the temperature and rate parameters of hot deformation on the mechanical behavior of forged titanium alloy VT14 under compression and tension. *Met. Sci. Heat Treat.* **2023**, *65*, 426–431. [CrossRef]
8. Ebied, S.; Hamada, A.; Gadelhaq, M.H.A.; Yamanaka, K.; Bian, H.-K.; Cui, Y.-J.; Chiba, A.; Gepreel, M.A.H. Study on hot deformation behavior of beta Ti-17Mo alloy for biomedical applications. *JOM* **2022**, *74*, 494–505. [CrossRef]
9. Li, H.-E.; Sha, A.-X. Effects of hot process parameters on flow stress and microstructures of TC18 titanium alloy. *J. Mater. Eng. Perform.* **2010**, *24*, 85–88.
10. Cui, J.-H.; Yang, H.; Sun, Z.-C. Research on hot deformation behavior and constitutive model of titanium alloy TB6. *Rare Met. Mat. Eng.* **2012**, *41*, 1166–1170.
11. Dan, D.-B.; Shi, K.; Xu, W.-C.; Lui, Y. Isothermal precision forging of Ti-6.5Al-3.5Mo-1.5Zr-0.3Si alloy impeller with twisted blades. *Trans. Nonferrous Met. Soc. China* **2006**, *016*, 2086–2090.
12. Mahadule, D.; Kumar, D.; Dandekar, T.R.; Khatirkar, R.K.; Suwas, S. Modelling of flow stresses during hot deformation of Ti-6Al-4Mo-1V-0.1Si alloy. *J. Mater. Res.* **2022**, *38*, 3750–3763. [CrossRef]
13. Roy, A.M.; Arróyave, R.; Sundararaghavan, V. Incorporating dynamic recrystallization into a crystal plasticity model for high-temperature deformation of Ti-6Al-4V. *Mater. Sci. Eng. A* **2023**, *880*, 145211. [CrossRef]
14. Sukumar, G.; Singh, B.B.; Balasundar, I.; Bhattacharjee, A.; Sarma, V.S. Primary hot working characteristics and microstructural evolution of as- cast and homogenized Ti-4Al-2.5V-1.5Fe alloy. *J. Alloys Compd.* **2023**, *947*, 169556. [CrossRef]
15. Jaimin, A.; Kotkunde, N.; Singh, S.K. Hot tensile rate-dependent deformation behaviour of AZ31B alloy using different Johnson-Cook constitutive models. *Arab. J. Sci. Eng.* **2023**, 1–16. [CrossRef]
16. Aranas, C.; Pasco, J.; McCarthy, T. Material constitutive modelling of Fe-based, Ni-based and high-entropy alloys: Development of Simu-Mat 1.0. *MRS Commun.* **2022**, *12*, 585–591. [CrossRef]
17. Olejarczyk-Wozenska, I.; Mrzygłód, B.; Hojny, M. Modelling the high-temperature deformation characteristics of S355 steel using artificial neural networks. *Arch. Civ. Mech. Eng.* **2022**, *23*, 1. [CrossRef]
18. Jaimin, A.; Kotkunde, N.; Singh, S.K.; Saxena, K.K. Studies on flow stress behaviour prediction of AZ31B alloy: Microstructural evolution and fracture mechanism. *J. Mater. Res. Technol.* **2023**, *27*, 5541–5558. [CrossRef]
19. Murugesan, M.; Yu, J.H.; Chung, W.J.; Lee, C.W. Warm deformation behavior and flow stress modeling of AZ31B magnesium alloy under tensile deformation. *Materials* **2023**, *16*, 5088. [CrossRef]
20. Singh, G.; Chakraborty, P.; Tiwari, V. Constitutive behavior of a homogenized at61 magnesium alloy under different strain rates and temperatures: An experimental and numerical investigation. *J. Mater. Civ. Eng.* **2023**, *35*, 04023314. [CrossRef]
21. Deb, S.; Muraleedharan, A.; Immanuel, R.J.; Panigrahi, S.K.; Racineux, G.; Marya, S. Establishing flow stress behaviour of Ti-6Al-4V alloy and development of constitutive models using Johnson-cook method and artificial neural network for quasi-static and dynamic loading. *Theor. Appl. Fract. Mech.* **2022**, *119*, 103338. [CrossRef]

22. Yang, H.-B.; Bu, H.-Y.; Li, M.-N.; Lu, X. Prediction of flow stress of annealed 7075Al alloy in hot deformation using strain-compensated arrhenius and neural network models. *Materials* **2021**, *14*, 5986. [CrossRef]
23. Yu, R.-H.; Li, X.; Li, W.-J.; Chen, J.-T.; Xin, G.; Li, J.-H. Application of four different models for predicting the high-temperature flow behavior of TG6 titanium alloy. *Mater. Today Commun.* **2021**, *26*, 102004. [CrossRef]
24. Rogoff, E.; Antony, M.; Markle, P. Calculating Ti-6Al-4V β transus through a chemistry based equation derived from combined element binary phase diagrams. *J. Mater. Eng. Perform.* **2018**, *27*, 5227–5235. [CrossRef]
25. Wanjara, P.; Jahazi, M.; Monajati, H.; Yue, S.; Immarigeon, J.P. Hot working behavior of near- alloy IMI834. *Mat. Sci. Eng. A* **2005**, *369*, 50–60. [CrossRef]
26. Semiatin, S.L.; Lahoti, G.D. Deformation and unstable flow in hot torsion of Ti-6AI-2Sn-4Zr-2 Mo-0.1Si. *Met. Mater. Trans. A* **1981**, *12*, 1719–1728. [CrossRef]
27. Wang, S.-S.; Deng, X.-Q.; Gao, P.-F.; Ren, Z.-P.; Wang, X.-X.; Feng, H.-L.; Zeng, L.-Y.; Zheng, Z.-D. Physical constitutive modelling of hot deformation of titanium matrix composites. *Int. J. Mech. Sci.* **2024**, *262*, 108712. [CrossRef]
28. Sellars, C.M.; Tegart, W.J. La relation entre le resistance et la structure dans la deformation a chaud. *Mem. Sci. Rev. Metall.* **1966**, *63*, 731–746.
29. Sun, Y.; Zeng, W.-D.; Zhao, Y.-Q. Development of constitutive relationship model of Ti600 alloy using artificial neural network. *Comp. Mater. Sci.* **2010**, *48*, 686–691. [CrossRef]
30. Long, J.-C.; Xia, Q.-X.; Xiao, G.; Qin, Y.; Yuan, S. Flow characterization of magnesium alloy ZK61 during hot deformation with improved constitutive equations and using activation energy maps. *Int. J. Mech. Sci.* **2021**, *191*, 106069. [CrossRef]
31. Shen, C.-W.; Yang, H.; Sun, Z.-C.; Cui, J.-H. Based on BP artificial neural network to building the constitutive relationship of TA15 alloy. *J. Plast. Eng.* **2007**, *14*, 101–104+132.
32. Stöcker, J.; Fuchs, A.; Leichsenring, F.; Kaliske, M. A novel self-adversarial training scheme for enhanced robustness of inelastic constitutive descriptions by neural networks. *Comput. Struct.* **2022**, *265*, 106774. [CrossRef]
33. Shaheen, M.A.; Presswood, R.; Afshan, S. Application of machine learning to predict the mechanical properties of high strength steel at elevated temperatures based on the chemical composition. *Structure* **2023**, *52*, 17–29. [CrossRef]
34. Mcclelland, J.; Rumelhart, D. *Parallel Distributed Processing;* Explorations in the Microstructure of Cognition: Psychological and Biological Models; MIT Press: Cambridge, MA, USA, 1987.
35. Murugesan, M.; Yu, J.H.; Chung, W.J.; Lee, C.W. Hybrid artificial neural network-based models to investigate deformation behavior of AZ31B magnesium alloy at warm tensile deformation. *Materials* **2023**, *16*, 5308. [CrossRef] [PubMed]
36. Ahmadi, H.; Ashtiani, H.R.R.; Heidari, M. A comparative study of phenomenological, physically-based and artificial neural network models to predict the hot flow behavior of API 5CT-L80 steel. *Mater. Today Commun.* **2020**, *25*, 101528. [CrossRef]

MDPI AG
Grosspeteranlage 5
4052 Basel
Switzerland
Tel.: +41 61 683 77 34

Materials Editorial Office
E-mail: materials@mdpi.com
www.mdpi.com/journal/materials

www.ingramcontent.com/pod-product-compliance
Lightning Source LLC
Chambersburg PA
CBHW061614210926
51850CB00009B/2322
* 9 7 8 3 7 2 5 8 6 0 6 1 6 *